AGRICULTURE ISSUES AND POLICIES

FARM PROGRAMS

PREVENTING PAYMENTS FOR EXCESSIVE INCOMES AND DECEASED INDIVIDUALS

AGRICULTURE ISSUES AND POLICIES

Additional books in this series can be found on Nova's website under the Series tab.

Additional E-books in this series can be found on Nova's website under the E-book tab.

AGRICULTURE ISSUES AND POLICIES

FARM PROGRAMS

PREVENTING PAYMENTS FOR EXCESSIVE INCOMES AND DECEASED INDIVIDUALS

EDWARD A. ROEDER
EDITOR

NOVINKA
New York

For permission to use material from this book please contact us:
Telephone 631-231-7269; Fax 631-231-8175
Web Site: http://www.novapublishers.com

Additional color graphics may be available in the e-book version of this book.

Library of Congress Cataloging-in-Publication Data

ISBN: 978-1-62948-622-2

Published by Nova Science Publishers, Inc. † New York

CONTENTS

PREFACE

The Government Accountability Office (GAO) was asked to review FSA's income verification practices. Another concern has been the distribution of benefits to ineligible participants, including potentially improper payments to deceased individuals, which, as the GAO and others have reported, may call into question whether these farm safety net programs are benefiting the agricultural sector as intended. This book examines the FSA's (1) review of tax returns and (2) review of accountants' and attorneys' statements and (3) FSA's and NRCS's recovery of payments to participants who exceeded income limits and (4) the extent to which procedures are in place to prevent the FSA and NRCS from making potentially improper payments to deceased individuals.

Chapter 1 - In light of high farm incomes and constrained federal budgets, the cost of federal farm and conservation programs—about $15 billion annually from 2009 through 2012—has come under scrutiny. Under the 2008 Farm Bill, participants whose incomes exceed specific limits are ineligible for certain program payments. USDA's FSA makes income eligibility determinations for programs it administers and also for conservation programs administered by NRCS. FSA verifies that participants have incomes below the limits by reviewing either tax returns (with consent from participants) or statements from accountants or attorneys. GAO was asked to review FSA's income verification practices. This report examines FSA's (1) review of tax returns and (2) review of accountants' and attorneys' statements and (3) FSA's and NRCS's recovery of payments to participants who exceeded income limits. GAO reviewed 115 tax return files and 163 files with accountants' and attorneys' statements from 18 FSA state offices selected to reflect geographic and program diversity, analyzed agency data, and interviewed agency officials.

Chapter 2 - USDA spends about $20 billion annually on federal programs that support farm income, conserve natural resources, and help farmers manage risks from natural disasters, benefiting over 1 million participants. Given their cost and continuing nationwide budget pressures, these programs have come under increasing scrutiny. One concern has been the distribution of benefits to ineligible participants, including potentially improper payments to deceased individuals, which, as GAO and others have reported, may call into question whether these farm safety net programs are benefiting the agricultural sector as intended. GAO was asked to evaluate USDA controls over payments to the deceased. This report examines the extent to which procedures are in place to prevent (1) FSA and (2) NRCS from making potentially improper payments to deceased individuals and (3) RMA from providing potentially improper subsidies on behalf of deceased individuals. GAO reviewed a random sample of payments, compared USDA's databases with SSA's master list of deceased individuals, and interviewed agency officials.

PREFACE

The Government Accountability Office (GAO) was asked to review FSA's income verification practices. Another concern has been the distribution of benefits to ineligible participants, including potentially improper payments to deceased individuals, which, as the GAO and others have reported, may call into question whether these farm safety net programs are benefiting the agricultural sector as intended. This book examines the FSA's (1) review of tax returns and (2) review of accountants' and attorneys' statements and (3) FSA's and NRCS's recovery of payments to participants who exceeded income limits and (4) the extent to which procedures are in place to prevent the FSA and NRCS from making potentially improper payments to deceased individuals.

Chapter 1 - In light of high farm incomes and constrained federal budgets, the cost of federal farm and conservation programs—about $15 billion annually from 2009 through 2012—has come under scrutiny. Under the 2008 Farm Bill, participants whose incomes exceed specific limits are ineligible for certain program payments. USDA's FSA makes income eligibility determinations for programs it administers and also for conservation programs administered by NRCS. FSA verifies that participants have incomes below the limits by reviewing either tax returns (with consent from participants) or statements from accountants or attorneys. GAO was asked to review FSA's income verification practices. This report examines FSA's (1) review of tax returns and (2) review of accountants' and attorneys' statements and (3) FSA's and NRCS's recovery of payments to participants who exceeded income limits. GAO reviewed 115 tax return files and 163 files with accountants' and attorneys' statements from 18 FSA state offices selected to reflect geographic and program diversity, analyzed agency data, and interviewed agency officials.

Chapter 2 - USDA spends about $20 billion annually on federal programs that support farm income, conserve natural resources, and help farmers manage risks from natural disasters, benefiting over 1 million participants. Given their cost and continuing nationwide budget pressures, these programs have come under increasing scrutiny. One concern has been the distribution of benefits to ineligible participants, including potentially improper payments to deceased individuals, which, as GAO and others have reported, may call into question whether these farm safety net programs are benefiting the agricultural sector as intended. GAO was asked to evaluate USDA controls over payments to the deceased. This report examines the extent to which procedures are in place to prevent (1) FSA and (2) NRCS from making potentially improper payments to deceased individuals and (3) RMA from providing potentially improper subsidies on behalf of deceased individuals. GAO reviewed a random sample of payments, compared USDA's databases with SSA's master list of deceased individuals, and interviewed agency officials.

In: Farm Programs
Editor: Edward A. Roeder
ISBN: 978-1-62948-622-2

Chapter 1

FARM PROGRAMS: ADDITIONAL STEPS NEEDED TO HELP PREVENT PAYMENTS TO PARTICIPANTS WHOSE INCOMES EXCEED LIMITS*

United States Government Accountability Office

WHY GAO DID THIS STUDY

In light of high farm incomes and constrained federal budgets, the cost of federal farm and conservation programs—about $15 billion annually from 2009 through 2012—has come under scrutiny. Under the 2008 Farm Bill, participants whose incomes exceed specific limits are ineligible for certain program payments. USDA's FSA makes income eligibility determinations for programs it administers and also for conservation programs administered by NRCS. FSA verifies that participants have incomes below the limits by reviewing either tax returns (with consent from participants) or statements from accountants or attorneys. GAO was asked to review FSA's income verification practices. This report examines FSA's (1) review of tax returns and (2) review of accountants' and attorneys' statements and (3) FSA's and NRCS's recovery of payments to participants who exceeded income limits.

* This is an edited, reformatted and augmented version of the Highlights of GAO-13-741, a report to the Chairwoman, Committee on Agriculture, Nutrition and Forestry, U.S. Senate, dated August 2013.

GAO reviewed 115 tax return files and 163 files with accountants' and attorneys' statements from 18 FSA state offices selected to reflect geographic and program diversity, analyzed agency data, and interviewed agency officials.

What GAO Recommends

To reduce the risk of improper payments to participants whose incomes exceed statutory limits, Congress should consider simplifying those limits. GAO recommends that FSA monitor state office reviews of tax returns and accountants' and attorneys' statements and implement a process to verify that these statements accurately reflect incomes. USDA generally agreed with GAO's findings and recommendations.

What GAO Found

As part of verifying if farm and conservation program participants had incomes below statutory limits—making them eligible to receive certain 2009 and 2010 program payments—reviews of tax returns by the U.S. Department of Agriculture's (USDA) Farm Service Agency's (FSA) state offices varied in quality. GAO's review of 115 tax return files from selected state offices found that some files met agency guidance and had no apparent errors. Other files did not meet agency guidance or contained errors, resulting in some potentially improper payments to participants whose incomes exceeded the limits. For example, GAO found errors in 19 of the 22 tax return files it reviewed from FSA offices in two states; one of these errors led to a potentially improper payment of $40,000. FSA headquarters does not monitor state offices' reviews of tax returns to ensure that the offices are applying program guidance consistently and making accurate eligibility determinations, even though federal standards for internal control direct agencies to monitor and assess the quality of performance over time. Also, 2008 Farm Bill provisions requiring a distinction between farm and nonfarm income make it difficult for agency officials to verify if participants' incomes exceed the limits without making errors. Because the statutory limits for farm and nonfarm income differ, to verify such income, FSA officials must comb through sometimes long and complex tax returns to classify and calculate income—a difficult task for those who are not accountants or tax preparers. Recent bills in the House and Senate have proposed using total adjusted gross income instead of farm and nonfarm income, which would reduce the need for FSA to review tax returns.

When relying on accountants' and attorneys' statements to verify participants' incomes for 2009 and 2010, FSA state offices sometimes accepted statements that did not meet agency guidance or contained errors, resulting in some questionable eligibility determinations and potential payments to participants whose income exceeded statutory limits. GAO's review of 163 files with accountants' and attorneys' statements from selected state offices found that some state offices followed FSA's guidance in full, but others sometimes did not. For example, 14 of the 16 statements GAO reviewed from one FSA state office met agency guidance, whereas 21 of the 39 statements GAO reviewed in two other state offices did not. In addition, some accountants' and attorneys' statements contained errors, such as miscalculations of average income. FSA's headquarters does not monitor its state offices to ensure that they accept only statements meeting agency guidance or verify the accuracy of participants' income in these statements by reviewing supporting documentation. As a result, FSA cannot be assured that the statements are accurate or that payments are being made only to participants whose incomes fall below statutory limits.

FSA and USDA's Natural Resources Conservation Service (NRCS) are each responsible for recovering any payments made by their respective programs to ineligible participants. In May 2012, FSA started to recover about $143 million in overpayments made to its participants in 2009 and 2010, but NRCS has not identified the amount of overpayments made or begun recovering payments it made to ineligible participants, because it had to first update project management software in February 2013. NRCS issued new guidance with procedures for identifying and collecting overpayments that were made and expects to send letters by September 2013 seeking reimbursement of overpayments.

ABBREVIATIONS

FSA	Farm Service Agency
IRS	Internal Revenue Service
NRCS	Natural Resources Conservation Service
USDA	U.S. Department of Agriculture

August 29, 2013

The Honorable Debbie Stabenow
Chairwoman
Committee on Agriculture, Nutrition and Forestry
United States Senate

Dear Madam Chairwoman:

Under federal farm and conservation programs from 2009 through 2012, about 1.4 million participants received about $15 billion annually in payments for crop subsidies, disaster assistance, and conservation practices.[1] During the same period, while most of the nation was in the midst of an economic downturn, overall farm income levels increased dramatically. The U.S. Department of Agriculture (USDA) reported that net farm income was $63 billion in 2009, generally rose through 2012, and is projected to reach a record $128 billion in 2013—more than double net farm income in 2009. In light of high farm incomes and rising federal budget deficits, the costs to the federal government of farm and conservation programs have come under heightened scrutiny. In particular, members of Congress and others have raised questions about whether some payments may be going to participants who are not eligible to receive them. By law, participants in certain farm and conservation programs are ineligible for payments if their incomes exceed specific limits. Within USDA, the Farm Service Agency (FSA)—through the agency's network of headquarters, state, and county offices—is responsible for administering the bulk of farm program payments, and the Natural Resources Conservation Service (NRCS) administers payments for most conservation programs.[2]

The Food, Conservation, and Energy Act of 2008 (2008 Farm Bill) establishes the current income limits for many farm and conservation programs. Separate limits apply to farm income and nonfarm income; both limits are based on adjusted gross income (or a comparable measure) as defined in the Internal Revenue Code and averaged over the 3 most recent tax years. Specifically, a participant is ineligible to receive payments under some farm programs if his or her average adjusted gross nonfarm income exceeds $500,000 and to receive another type of payment if his or her average adjusted gross farm income exceeds $750,000.[3] A participant is ineligible to receive conservation payments if his or her average adjusted gross nonfarm income exceeds $1 million, unless at least 66.66 percent of his or her average adjusted gross income is derived from average adjusted gross farm income.[4]

In October 2008, we reported that from 2003 through 2006, FSA paid nearly $50 million under farm programs to participants whose incomes may have exceeded the income limit at that time ($2.5 million annually per individual or entity), potentially making them ineligible.[5] These payments occurred primarily because FSA did not have management controls, such as reviews of an appropriate sample of recipients' tax returns, to verify that payments were going only to participants who did not exceed the income limit. We recommended that FSA work with the Internal Revenue Service (IRS) to develop a system to identify participants potentially exceeding income limits. In response to our recommendation, FSA and IRS implemented an income verification process in 2009. Under the new process, IRS identifies participants potentially exceeding the income limits, and FSA reviews information submitted by the participants to make eligibility determinations. Specifically, participants submit to FSA for review either (1) tax returns or (2) accountants' or attorneys' statements certifying the participants' income levels. For both FSA and NRCS programs, FSA is responsible for determining whether participants have incomes exceeding the limits.[6] Once FSA makes a determination, FSA and NRCS are each responsible for recovering any payments made to ineligible participants by their respective programs.

In this context, you asked us to assess FSA's implementation of a process to verify that participants' incomes complied with statutory limits for receiving farm and conservation payments. This report examines (1) FSA's review of participants' tax returns to verify that participants receiving payments do not exceed income limits, (2) FSA's review of accountants' and attorneys' statements to verify participants' incomes, and (3) FSA's and NRCS's recovery of overpayments to participants determined to have exceeded income limits.

To address these objectives, we reviewed FSA and NRCS files, analyzed agency data, interviewed agency officials, and reviewed and updated our past work. Specifically, to examine FSA's review of participants' tax returns and accountants' and attorneys' statements to verify participants' incomes for 2009 and 2010 (the most recent years for which data were available), we reviewed 278 FSA files from 18 FSA state offices. We first selected a nonprobability sample of 18 state offices to reflect diversity with respect to number of income eligibility reviews conducted by the office, size and type of farm and conservation programs, and geographic location. Eight of these 18 state offices had reviewed 30 or more files. For these 8, we drew a random sample of 20 to

30 files from each office for our review. The other 10 state offices in our sample had reviewed fewer than 30 files, so we selected all of them for our review. In each of our original random samples, half the files contained tax returns, half contained accountants' and attorneys' statements, and both groups contained determinations of both eligibility and ineligibility. Because this random sample of files was based on preliminary data from the agency, the proportions of files with tax returns versus accountants' and attorneys' statements changed during our work; in all, we reviewed 115 files that contained tax returns and 163 files that contained accountants' and attorneys' statements.[7] We reviewed the files to evaluate the state offices' compliance with agency guidance and to understand the offices' rationale for their eligibility determinations. Our sampling methods do not allow our results to be generalized to all 51 FSA state offices nationwide or to a single state office (except in cases where we selected all the files). Nevertheless, by providing examples, the information from these files enhanced our understanding of FSA's income verification process.

In addition to reviewing files from selected FSA state offices, we collected nationwide FSA data on the agency's process for determining income eligibility. To assess the reliability of these data, we (1) performed electronic testing of required data elements, (2) reviewed existing information about the data and the system that produced them, and (3) interviewed agency officials knowledgeable about the data. We found these data to be sufficiently reliable for purposes of this report. We analyzed FSA's data to identify any differences among state offices in rates of determinations of eligibility. We also interviewed FSA headquarters officials and officials in eight state offices, which had reviewed more than 30 files, to understand the agency's policies, procedures, and practices for determining participants' eligibility. To examine FSA's recovery of overpayments to participants determined to have exceeded income limits in 2009 and 2010, we collected nationwide FSA data on payment amounts to be recovered and interviewed agency officials about policies, procedures, and practices used to recover such payments. We again assessed the reliability of the data through a review of the system that generated the data, as well as other steps, and determined that they were reliable for purposes of this report. To examine NRCS's payment recovery process, we reviewed agency guidance and relevant reports by the USDA Office of Inspector General, and we interviewed agency officials in headquarters and state offices. Appendix I describes our objectives, scope, and methodology in more detail.

We conducted this performance audit from June 2012 to August 2013 in accordance with generally accepted government auditing standards. Those standards require that we plan and perform the audit to obtain sufficient, appropriate evidence to provide a reasonable basis for our findings and conclusions based on our audit objectives. We believe that the evidence obtained provides a reasonable basis for our findings and conclusions based on our audit objectives.

BACKGROUND

A variety of federal farm programs provide benefits to participants, including payments, to help protect against the risks of low crop prices or bad weather, among other hardships. These payments are made in accordance with individual program rules to participants who own agricultural land or produce certain crops. In addition, conservation payments provide assistance to participants to help them safeguard environmentally sensitive land by, for example, retiring land from agriculture or implementing practices such as erosion control measures that protect the land during farming. Participants can receive these payments directly or through legal entities including partnerships, corporations, and trusts.

The 2008 Farm Bill established eligibility rules for farm and conservation programs, including separate income limits for an individual's or legal entity's farm income and nonfarm income. These income limits remain in effect through September 2013. Both types of income use as their foundation total adjusted gross income, as defined in the Internal Revenue Code, or a comparable measure,[8] and both types are averaged over the 3 most recent tax years. Under present limits, participants are not eligible to receive some farm payments if their average adjusted gross nonfarm income exceeds $500,000; another type of farm payment if their average adjusted gross farm income exceeds $750,000; and conservation payments if their average adjusted gross nonfarm income exceeds $1 million, unless at least 66.66 percent of their average adjusted gross income is average adjusted gross farm income. Because these income limits apply to individuals, under certain conditions, a husband and wife could collectively earn up to $1 million in average adjusted gross nonfarm income and $1.5 million in average adjusted gross farm income and be eligible for most farm payments, or up to $2 million in average adjusted gross nonfarm income and be eligible for conservation payments. The 2008 Farm Bill specifies that certified public accountants (or others designated by

USDA) may provide a statement certifying an allocation of income between individuals who file a joint tax return, as if they had filed separately. The 2008 Farm Bill also allows USDA to waive the income limit for conservation payments in cases involving "environmentally sensitive land of special significance."[9] According to NRCS documents, the agency waived the income limit 4 times in 2009 and 15 times in 2010, to approve conservation contracts for about $89 million in total to be paid over multiple years. FSA also waived the income limit 12 times in 2010, to approve conservation contracts together granting about $80,000 to be paid in 2010 and additional amounts to be paid over multiple years.[10] See appendix II for a more detailed description of the 2008 Farm Bill's income limits and appendix III for more details about NRCS's and FSA's waivers.

The 2008 Farm Bill directs the Secretary of Agriculture to include income related to the following categories, among others, when determining average adjusted gross farm income:

- production, feeding, and rearing of livestock;
- production of products derived from or produced by livestock;
- production of crops and unfinished raw forestry products;
- processing, packing, storing, and transporting of farm, ranch, and forestry commodities, including renewable energy;
- sale of land that has been used for agriculture;
- sale of easements of farmland, ranchland, or forestry land; water or hunting rights; or environmental benefits;
- rental or lease of land or equipment used for farming, ranching, or
- forestry operations, including water or hunting rights;
- production of farm-based renewable energy; and
- sale of equipment to conduct farm, ranch, or forestry operations, if average adjusted gross farm income is at least two-thirds of a participant's average adjusted gross income.

Nonfarm income is defined as total income minus farm income. Appendix II also includes a more detailed description of the 2008 Farm Bill's definitions of farm income and nonfarm income.

In October 2008, we reported that FSA could not ensure that all individuals receiving farm payments had incomes below eligibility limits because the agency's management controls were not targeted to provide such assurance.[11] From 2003 through 2008, to ensure that only eligible individuals received farm payments, either directly or as members of entities, FSA relied on participants' one-time self-certifications that they met income eligibility requirements and the participants' promises to notify FSA if they no longer met these requirements. As we reported, however, these self-certifications did not always prove sufficient, and FSA's principal management control—a review of a sample of program participants to verify compliance with income limits—did not routinely target high-income participants. In the 2008 Farm Bill, Congress directed the Secretary of Agriculture to establish statistically valid procedures to conduct targeted audits of participants the Secretary determines to be most likely to exceed the legislation's income limits. Consistent with this provision and our findings, we recommended that FSA work with IRS to develop a system for verifying income eligibility for all recipients of farm payments.

In 2009, USDA and IRS signed a memorandum of understanding, launching for the first time a process in which IRS screens participants initially to identify those whose incomes may exceed limits, and FSA makes an eligibility determination, as shown in figure 1.

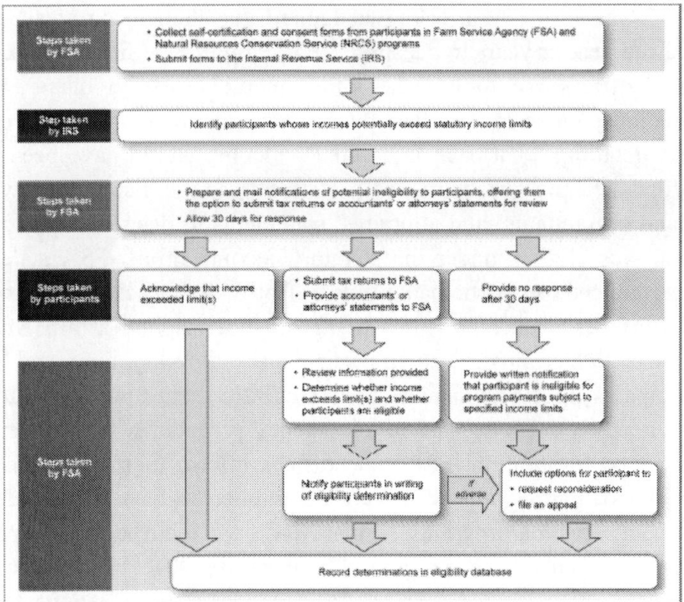

GAO and Farm Service Agency.

Figure 1. Farm Service Agency's (FSA) Process for Determining If Participants' Incomes Exceed Statutory Limits.

IRS and FSA implemented this process for 2009 and 2010, and in March 2013, FSA began the process for 2011 and 2012. Specifically, the process starts when participants provide consent for IRS to disclose certain tax-related information to FSA to verify that they meet applicable income limits.[12] For 2009 and 2010, farm and conservation program participants were to complete a self-certification form saying they met income limits and a consent form for IRS to release limited tax information; for participants who completed both forms, IRS used its tax database to estimate farm income and nonfarm income according to USDA instructions. IRS computer programs compared these income estimates against the 2008 Farm Bill's income limits to identify participants who may have exceeded these limits, and IRS provided the resulting list to FSA.[13] The agency then notified potentially ineligible participants by mail and gave them the opportunity to provide documentation if they believed their income did not exceed the eligibility limits. According to FSA's handbook on determining eligibility for farm payments, participants can provide to FSA state offices (1) tax returns for the 3 most recent years, (2) accountants' or attorneys' statements certifying the participants' income levels for these 3 years, or (3) an acknowledgment that their income exceeded the limits for the year or years in question.[14] For couples who file joint tax returns, FSA state offices are to use the joint income levels to make eligibility determinations, unless a certified public accountant or attorney provides a statement detailing what each individual's income would have been had the couple filed separate tax returns. FSA state offices are then to review the tax returns and accountants' and attorneys' statements to determine whether each individual was eligible under the statutory income limits. FSA state offices deem ineligible those participants who (1) provide an acknowledgment that their incomes exceeded the limits or (2) do not respond at all. After state offices make a determination about participants' eligibility, they send letters to the participants notifying them of their eligibility, enter the eligibility status into the agency's database, and follow agency procedures for recovering any overpayments. In addition, FSA state offices inform their state-level NRCS counterparts of participants who were determined to have exceeded income limits for conservation programs, so that NRCS can recover any overpayments made to participants in its programs. In February 2013, USDA's Office of Inspector General reported that NRCS needed to strengthen its efforts to ensure that program benefits reach only eligible participants and serve their intended purposes.[15] Moreover, the Inspector General found that NRCS has had long-standing problems with verifying participants' eligibility.

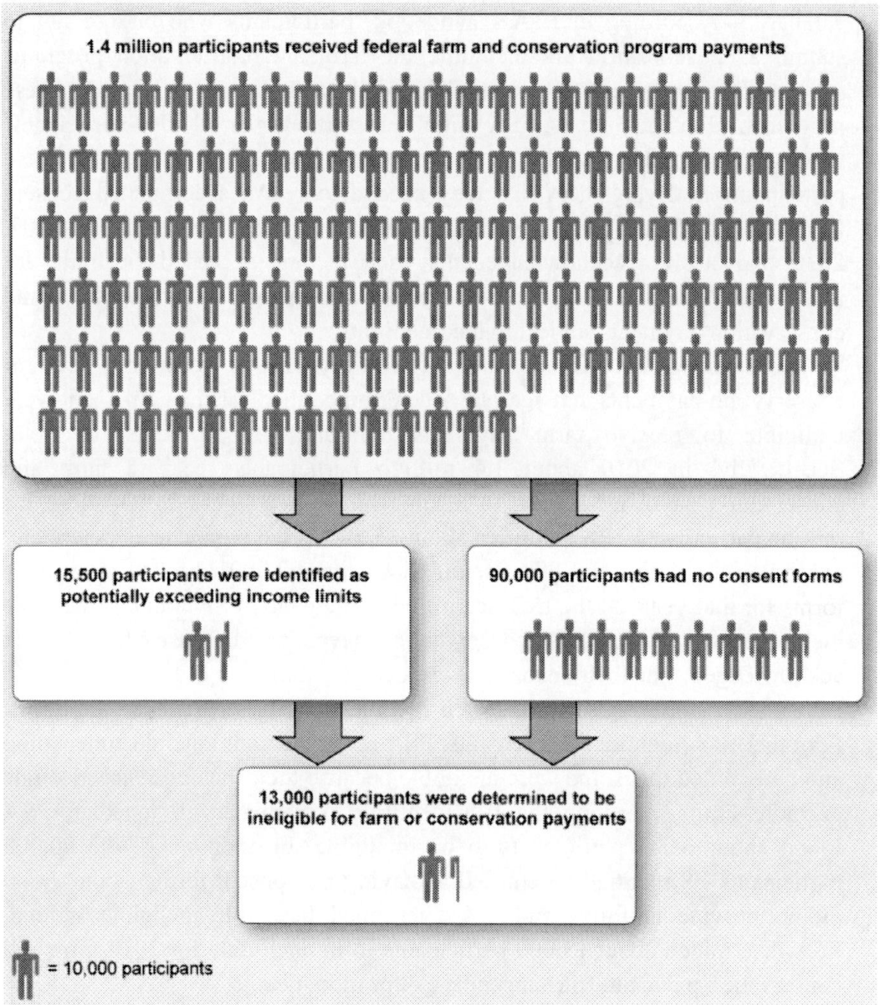

GAO analysis of Farm Service Agency data.

Figure 2. Fraction of Participants Receiving Farm and Conservation Payments in 2010 Whose Incomes Exceeded Limits.

For 2009, 2010, and 2011, FSA also worked with IRS to identify participants who (1) had received farm or conservation payments but had not completed a consent form allowing IRS to share limited information with FSA or (2) had completed a consent form but had had the form rejected because, for example, the name on the form did not precisely match the name in IRS's database.[16] According to FSA's handbook, participants who choose not to submit a consent form are ineligible for farm and conservation programs subject to income limits and must refund all payments received under these programs.[17] For the 3 years, FSA initially identified over 400,000 participants lacking consent forms. FSA state and county offices sent letters to these participants notifying them that they needed to return a completed consent form. About 28,000 participants either did not provide the forms for 2009, 2010, and 2011 or acknowledged that their income exceeded the limits for those years, and FSA determined them to be ineligible for the farm and conservation payments subject to income limits.

Considering 2010 alone, out of all participants receiving farm or conservation payments that year, FSA determined that less than 1 percent were ineligible to receive farm payments, conservation payments, or both. Specifically, in 2010 about 1.4 million participants received farm and conservation payments subject to income limits (see figure 2). Of these 1.4 million participants, IRS identified about 15,500 as having incomes potentially exceeding the limits, and FSA identified another 90,000 who had no consent forms for that year. Of the IRS- identified participants, FSA found some to be ineligible for farm and conservation payments because they either acknowledged that their incomes exceeded the limits or did not respond to FSA's letter notifying them that their incomes may have exceeded the limits. For other IRS- identified participants, FSA reviewed additional documentation and determined that some were ineligible because their incomes exceeded one or more limits.[18] Among the IRS-identified participants all together, FSA found about 6,000 participants to be ineligible. In addition, of the 90,000 participants FSA initially identified as having no consent forms, about 7,000 did not provide the forms, and FSA determined them to be ineligible. In total, FSA determined about 13,000 participants to be ineligible for 2010 farm and conservation payments. See appendix IV for more details.

Congress has begun deliberations over the next farm bill, and proposals revising farm and conservation program income limits have been made. In particular, in June 2013, the Senate passed one version of a farm bill— the Agriculture Reform, Food, and Jobs Act of 2013 (S. 954)—and in July 2013, the House of Representatives approved another version—the Federal Agriculture Reform and Risk Management Act of 2013 (H.R. 2642)—both with income limit provisions. Although neither bill has become law, both bills provide further insights into congressional deliberations over the next farm bill. For example, both bills eliminate the distinction between farm income and nonfarm income and establish a single limit on total adjusted gross income for most farm programs, simplifying implementation of the income limits as a result. Simplifying the income limits might affect the universe of participants eligible for farm and conservation payments.[19] The Senate bill set a limit of $750,000, and the House bill set a limit of $950,000. As with current income limits, these limits would apply to individuals, so under certain conditions a husband and wife could collectively have an income of up to $1.5 million under the Senate bill or up to $1.9 million under the House bill. In addition, the Senate bill included a provision that would reduce subsidies for crop insurance premiums by 15 percentage points for participants with average adjusted gross income over $750,000.[20]

FSA State Offices' Reviews of Tax Returns Vary in Quality and Have Led to Some Erroneous Eligibility Determinations

When determining if participants' incomes fell below the limits, qualifying them for 2009 and 2010 farm payments, FSA state offices did not always review tax returns according to agency guidance and sometimes made errors when calculating and classifying income, or they relied on insufficient documentation. As a result, FSA state offices may have made some erroneous determinations of participants' eligibility or ineligibility, resulting in some potentially improper payments to ineligible participants whose incomes exceeded the limits. The absence of FSA monitoring of state offices' reviews, as well as complex income eligibility policies, both contribute to variation in quality of reviews performed by state offices.

FSA State Offices Did Not Always Follow Guidance and Sometimes Made Errors When Reviewing Participants' Tax Returns

To help ensure accuracy of payments, FSA developed new guidance in its handbook on how to determine eligibility for farm payments, detailing procedures its state offices are to follow when reviewing tax returns to verify participants' incomes. According to the handbook, for each participant under review in 2009 and 2010, state offices were to maintain a file containing all of the following documents:

- copies of complete tax returns received from the participant for all relevant years;
- printout showing eligibility status recorded in FSA's database before state office review;
- copies of the original self-certification forms submitted by the participant;
- calculation worksheets for average farm income and nonfarm income for 2009, 2010, or both years;
- printout showing updated eligibility status recorded in FSA's database after state office review, if changed; and
- copy of the letter notifying the participant of the agency's final eligibility determination.

The handbook also includes guidance about where to find and how to calculate total income, farm income, and nonfarm income using various IRS tax forms. For farm income, the handbook lists some of the IRS schedules, forms, and line numbers where farm income may be reported, but the list is not comprehensive because reporting farm income and preparing tax returns may be done in many acceptable ways. The handbook also includes information on the definition of farm income, so officials can discern which items to classify as farm income, even if the items are not reported on the specified forms and on line numbers listed in the handbook. To help FSA officials calculate total and average farm income and nonfarm income for the 3 most recent tax years, the handbook includes a step-by-step worksheet where officials can enter dollar amounts from tax returns for each of the 3 years, then perform calculations as directed, and ultimately arrive at an eligibility determination for each applicable income limit. In addition to the handbook, for FSA's 2009 and 2010 income verification process, headquarters officials provided training briefings to state officials before they began reviewing tax returns, to help ensure that state officials understood and followed agency guidance.

In our reviews of 115 tax return files maintained by 13 selected FSA state offices for 2009 and 2010, we found many files that were complete or nearly complete, with clear documentation of the rationale behind eligibility determinations and no apparent errors. For example, 24 of the 25 files we reviewed from the California and Texas state offices contained the self-certification forms, calculation worksheets, and notification letters specified in guidance. In addition, 11 of the 12 tax return files we reviewed from the Arkansas state office were nearly complete and included thorough records—for example, including e-mail exchanges with headquarters officials to verify proper classification of farm income in unusual cases. Further, some state offices, including those in Arkansas and Texas, developed their own, more detailed calculation worksheets to track and sum individual farm income amounts reported on specific schedules and forms for each year, rather than record only the total annual amounts, as indicated on the handbook's worksheet. The more detailed worksheets furnished a transparent record of which income sources officials classified as farm income or nonfarm income each year and of how they calculated average farm income and nonfarm income for the 3-year period—a clear picture of the steps officials took to determine participants' eligibility.

In contrast, files from some state offices were routinely missing some of the documentation specified in FSA's handbook. For example, 39 of the 41 tax return files we reviewed from the Indiana, Kansas, and Missouri state offices were missing the original self-certification forms provided by participants, and 38 of the 41 files were missing printouts showing that the eligibility determination had been correctly entered into FSA's database. Also, 8 of the 9 files we reviewed from the Louisiana state office were missing calculation worksheets. Without all the specified documents—including the printout showing the eligibility determination in FSA's database—FSA cannot be assured that all of its state offices' determinations of participants' eligibility are correct and accurately recorded. Accurate recording of participants' eligibility status in the agency's database is of particular importance because it is the first step in recovering payments made to ineligible participants; if the status of ineligible participants is not accurately recorded, improper payments may not be identified or recovered.

In addition, some tax return files we reviewed also included FSA state officials' errors in classifying and calculating farm income and nonfarm income, raising questions about FSA's confidence that participants receiving payments do not exceed income limits. For example, in some files FSA state officials incorrectly counted income from dividends or the sale of a house as farm income or incorrectly counted income from the sale of livestock, farmland, or timber as nonfarm income. In other files, FSA state officials used income from the wrong line number on an IRS form (e.g., the line listing gross rather than net income), made errors calculating average farm income and average nonfarm income, or calculated income amounts that we could not verify. In FSA's California office, we found potential errors or misclassifications in 9 of the 11 tax return files we reviewed, and in the Indiana office, we found such issues in 10 of the 11 files we reviewed. For example, in 1 file FSA state officials incorrectly counted a sale of farmland as nonfarm income and erroneously determined the participant to be eligible for farm payments when he should have been ineligible. In another file involving a farming corporation under review, FSA state officials properly classified income from the corporation's farm operation as farm income in 2 of the 3 years under review but classified income from the same source erroneously as nonfarm income in the third year. As a result, the corporation received about $40,000 in 2010 farm payments that it should not have.

Moreover, FSA state officials sometimes counted income as farm income without asking for corroborating information, leading to potential errors. For example, officials could not always tell from a tax return whether a parcel of sold land had been used for agriculture and whether gains from the sale should therefore be counted as farm income. In some cases, officials counted sales of such land as farm income (e.g., some officials told us they did so if the parcel's address was in a predominantly rural county or if the reviewing official knew about the participant's circumstances). In other cases, absent explicit evidence on the tax return that a parcel had been used for agricultural purposes, officials said they counted such sales as nonfarm income, and in still other cases, the officials said they required documentation of the parcel's use before deciding. Neither the FSA handbook nor the training provided by agency headquarters addresses how state offices should handle this issue, but according to a headquarters official, a sale of land should be classified as nonfarm income unless the tax return includes explicit evidence that the sold land is farmland or the participant provides documentation showing the property's classification as farmland.

Further, in some files we reviewed, FSA state officials allocated joint income to individuals, even though neither the 2008 Farm Bill nor agency guidance directs the agency to do so. For example, in the Indiana state office, agency officials allocated joint income to individuals, determining one or both individuals to be eligible for farm payments, in eight of the nine files we reviewed in which FSA made decisions on the basis of joint tax returns. As a result, these individuals received payments in 2009 and 2010, totaling about $56,000, which they would not have received if state officials had based their determinations on the income reported in participants' tax returns, as called for in the handbook. In one California file we reviewed, FSA state officials also allocated joint income to individuals, resulting in payments to one individual who would not otherwise have received the payments.

In interviews, FSA state officials described taking different approaches to reviewing tax returns and, in some instances, handling similar cases in different ways. For example, some state offices delegated reviewing tax returns to officials who had experience in the area, such as farm loan specialists, and in other state offices, officials who had little or no experience with tax returns performed the reviews. Also, in some states, FSA officials said they reviewed every page of each tax return they received, and in other states, agency officials said they did not always review a participant's entire tax return for each year. Instead, in cases where a participant's nonfarm income had potentially exceeded the limit, some agency officials reviewed the tax returns page by page until they had classified enough income as farm income to bring the participant's nonfarm income below the limit and did not review the remaining pages because, the officials said, they had limited time and resources. The remaining, unreviewed pages, however, could have contained income information that might have affected the participant's eligibility. Officials in one state office said they counted as farm income only the amounts reported on the specific schedules and line numbers listed in FSA's handbook. In another state office, officials said they usually reviewed the first page of each year's tax return and did not look further unless something caught their attention. Agency officials in three states said that, for each tax return file, one official reviewed the documents and made an initial determination of the participant's eligibility, and a second official reviewed the file again to verify the accuracy of the determination. Officials in other states said a single person reviewed each file because of limited staff.

In addition, our analysis of data that FSA maintains for all state offices during the income verification process—including data on the determinations of the 27 state offices with more than 100 reviews of participants' eligibility to receive farm payments for 2010—identified variation across state offices in the percentage of participants determined to be eligible (see figure 3). In particular, FSA tracks the number of participants identified by IRS as potentially exceeding income limits and, of these, the number determined eligible or ineligible after review by state offices. According to eligibility data for 2010, the percentage of participants that state offices determined to be eligible ranged from 26 percent in Ohio to 79 percent in Arizona. It is unclear, however, whether the large variation across state offices in these percentages stems from variation in the quality of the state offices' reviews or from something else. A reason for this uncertainty is that FSA officials were not able to determine the extent to which variation in quality of reviews contributed to the different percentages of participants that state offices deemed eligible. For more details on FSA's eligibility data and percentages of participants determined to be eligible, by state office, see appendix V.

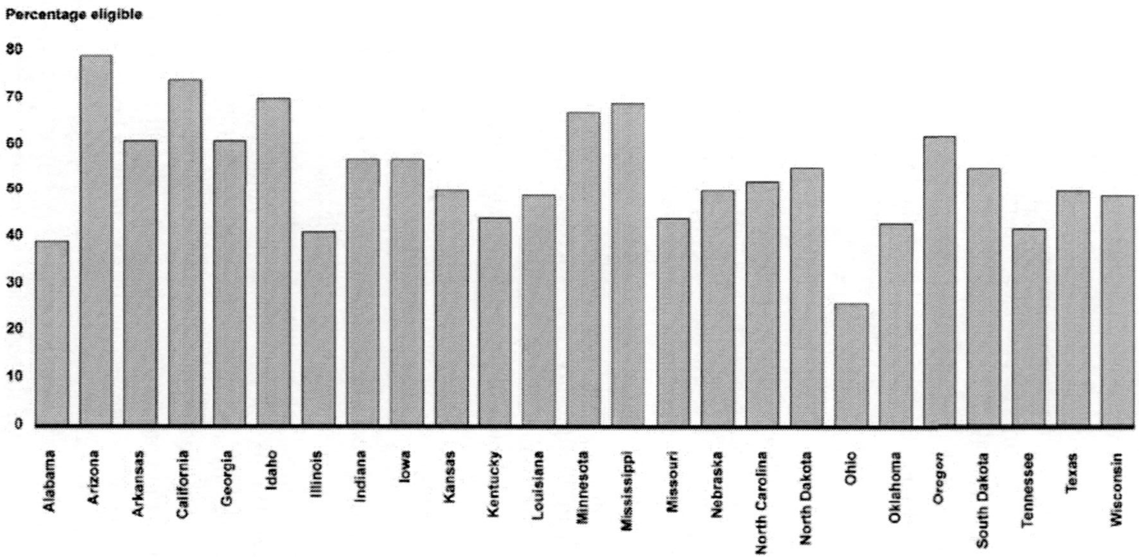

GAO analysis of Farm Service Agency data.

Note: Figure includes only states with more than 100 total reviews of participants' eligibility by income for 2010 farm program direct payments, which are fixed annual payments based on farms' historical crop production and accounted for the largest dollar amounts that year. FSA did not make a determination of participants' eligibility for these payments in all cases—for example, because some participants received no direct payments. We calculated percentages of participants determined to be eligible using only those cases in which FSA made a determination of eligibility for direct payments.

Figure 3. Percentage of Farm Program Participants Whose Files Were Reviewed by Selected Farm Service Agency (FSA) State Offices and Who Were Determined to Be Eligible for Certain 2010 Payments.

FSA Headquarters Does Not Assess the Quality of State Offices' Reviews, and Farm Bill Provisions Make Review of Tax Returns Difficult

FSA headquarters does not monitor state office reviews of tax returns to ensure that the offices are applying program guidance consistently and making accurate eligibility determinations on the basis of participants' incomes. One of the federal standards for internal control directs agencies to conduct monitoring to assess the quality of performance over time and ensure that any deficiencies are promptly resolved,[21] and regular monitoring of the quality and completeness of state offices' reviews can help prevent payments to participants who exceed income limits. Such monitoring would allow FSA to identify problem areas (e.g., if some state offices consistently misapply guidance), implement improvements to address problems, and help hold state offices accountable for accurate determinations of participants' eligibility to receive payments. Among other effects, such monitoring can help FSA assess whether the variation across state offices in the percentage of participants determined eligible for payments was related to the quality of state office reviews.

In addition, 2008 Farm Bill provisions establishing income limits can contribute to FSA state officials' making erroneous determinations of participants' eligibility to receive farm and conservation payments because the complexity of the provisions makes it difficult for the officials to verify if participants' incomes exceed the limits. FSA state officials in several states told us they felt ill-prepared to review tax returns because they are not accountants or tax preparers, and the 2008 Farm Bill's statutory limits further complicate the task. Specifically, because the eligibility limits for farm income and nonfarm income differ, the only way for FSA officials to verify that participants' incomes do not exceed the limits is to comb through sometimes long and complex tax returns to classify and calculate farm and nonfarm income—a difficult task for agency officials whose expertise is in farm programs, not accounting or tax preparation. FSA's handbook and training provide some basic guidance, but they cannot substitute for training as an accountant. Performing the task correctly requires not only a detailed understanding of the statutory income limits but also knowledge of various tax forms, business arrangements, and reporting options, and the difficulty of the task leaves FSA vulnerable to errors.

The complexity of verifying that participants' incomes do not exceed the limits is compounded by the 2008 Farm Bill's definition of farm income, which is itself complex. Specifically, the 2008 Farm Bill directs the Secretary of Agriculture to determine average adjusted gross farm income on the basis of a detailed list of categories, which must be interpreted and applied by FSA state officials when reviewing tax returns. To identify farm income reported on tax returns, state officials cannot rely only on the IRS schedules and lines explicitly labeled as farm income because the Farm Bill's definition is broader than IRS's definition. For example, the 2008 Farm Bill defines income from the sale or rental of farmland as farm income, but IRS generally does not. Consequently, FSA officials must know where to find income from land sales or rentals on IRS forms and, for each sale or rental, ascertain whether the land in question was used for agricultural purposes. Furthermore, FSA officials must interpret the detailed definition to identify subtle distinctions. For example, if a participant leases farmland for an oil or gas pipeline right-of-way, lease income is to be counted as farm income, but if a participant receives income from extraction of oil or gas from beneath farmland, this income is *not* to be counted as farm income. And if a participant receives income through certain types of farming partnerships or corporations, it is to be considered farm income, but if the participant receives dividends as a shareholder in other types of farming corporations, the dividends are to be classified as nonfarm income.[22] Also under the 2008 Farm Bill, the rental of farm equipment is to be counted as farm income in all cases, but the sale of farm equipment is to be considered farm income only if two- thirds of a participant's total income is farm income.

Recent bills in the House and Senate have proposed eliminating the distinction between farm income and nonfarm income, in favor of a limit on total adjusted gross income for farm and conservation program participants. Such income generally appears on a standard line or pair of lines on tax forms and could likely be verified through IRS's automated computer programs, with no need for FSA to conduct additional reviews in most cases, thereby simplifying FSA's task and offering greater assurance that farm payments would be made only to eligible participants. Without such simplification, FSA is likely to continue finding it difficult to accurately classify farm income and nonfarm income when reviewing tax returns. As a result, the agency may continue to make erroneous determinations of participants' eligibility, leading to potentially improper payments.

FSA'S REVIEW OF ACCOUNTANTS' AND ATTORNEYS' STATEMENTS CANNOT ENSURE THAT PARTICIPANTS RECEIVING PAYMENTS COMPLY WITH INCOME LIMITS

When relying on accountants' and attorneys' statements to determine if participants' incomes exceeded limits for 2009 and 2010, FSA state offices sometimes accepted statements that did not meet agency guidance, that contained errors, or both, resulting in some questionable eligibility determinations and potential payments to participants whose incomes exceeded statutory limits. FSA headquarters does not monitor its state offices' reviews of accountants' and attorneys' statements to ensure that they apply agency guidance consistently or verify the statements against supporting documentation to ensure accuracy.

FSA State Offices Sometimes Accepted Accountants' and Attorneys' Statements That Did Not Meet Agency Guidance or Contained Errors

FSA's handbook provides guidance for state officials when reviewing accountants' and attorneys' statements. State officials are to keep in each file a certain minimum set of documents related to eligibility determination. For files with accountants' or attorneys' statements, officials are to retain the statement and any attachments or enclosures. In addition, they are to retain many of the same documents as those required for tax return files, such as printouts showing the eligibility status recorded in FSA's database and a calculation worksheet for average farm income and nonfarm income. Also, according to FSA's handbook, state officials are to review the statements for certain key elements, including the following:[23]

- the accountant's or attorney's license number;
- acknowledgment that the accountant or attorney has reviewed and is familiar with, among other items, farm income and nonfarm income limits;
- the amount of annual farm income and nonfarm income for each relevant year;
- names of the IRS tax schedules and line numbers where farm income was reported on tax returns;

- average farm income and nonfarm income; and, if these averages appear to exceed the limits, a detailed explanation of the reason that the participant is eligible even so.

In our reviews of 163 files with accountants' and attorneys' statements from 18 selected state offices, we found many instances where FSA state officials kept in the files all the documents specified in agency guidance and accepted statements that included all key elements in the guidance. For example, in the Texas office, 14 of the 16 files we reviewed with accountants' or attorneys' statements included all the specified documents, and 14 of the 16 statements we reviewed included all the key elements. Further, several files from the California and Texas state offices included correspondence from the state office to accountants or attorneys, rejecting statements that did not include all elements specified in agency guidance and sometimes detailing what was needed, demonstrating these offices' firm adherence to agency guidance. And FSA officials in two state offices went beyond what was required by agency guidance to verify that accountants' and attorneys' licenses were valid and current.

Our reviews, however, also found files in which FSA state officials accepted statements of participants' incomes even though the statements did not contain all key elements specified in guidance. For example, of the accountants' and attorneys' statements we reviewed for California and Indiana, 6 of 15 and 15 of 24, respectively, were missing one or more key elements. Missing elements included income amounts and averages for all the relevant years, the names of the IRS schedules and line numbers where farm income was reported on tax returns, and a detailed explanation of why a participant's income did not exceed limits when information on tax returns indicated otherwise.

Regarding the detailed explanation, accountants' and attorneys' statements we reviewed varied, particularly when allocating income reported on joint tax returns to individuals. Some statements simply reported the amount of farm income and nonfarm income that was to be attributed to an individual participant, and it was not clear from the statement whether the participant had filed a joint tax return. Other accountants and attorneys explained that they had allocated joint income as if the participant had filed a separate tax return, and still others provided a detailed table showing how they allocated every increment of income to each of the individuals listed on a joint return and whether they classified each increment as farm income or nonfarm income. According to FSA's handbook, a detailed explanation is needed when the information on tax returns indicates that a participant's income exceeds one or more of the limits for farm income and nonfarm income, but the handbook does not specify whether an explanation is always needed for participants who file joint tax returns or what is needed in such an explanation. According to our review, FSA state offices did not interpret or apply this guidance consistently, sometimes rejecting statements that lacked a sufficient explanation, as directed by the handbook, and sometimes accepting statements that had no explanation.

When accountants and attorneys included an explanation of methodology or other key elements in their statements, the elements sometimes revealed errors, such as using the wrong years' income to calculate average incomes; counting some tax deductions twice; or using gross, rather than net, income to calculate totals—thereby reinforcing the importance of FSA's ensuring that statements contain the elements listed in the agency's guidance. Some FSA state officials recognized such errors and did not accept the statements, and other FSA state officials accepted statements with the errors. In one file we reviewed, an accountant included the names of IRS schedules and line numbers where farm income was reported, as called for in FSA's handbook; the information showed that the accountant had misclassified certain tax deductions as farm income, which brought the participant's nonfarm income below the limit. FSA's state office accepted the statement with the classification error and mistakenly determined that the participant was eligible to receive farm payments totaling more than $90,000 in 2009 and 2010. Further, we found a few files where FSA state officials introduced errors in calculations done as part of their reviews even though the accountants' or attorneys' statements had reported income correctly. For example, when calculating whether one participant's income exceeded the limits, a state office used nonfarm income in place of total income and mistakenly found the participant to be eligible for 2010 farm payments.

Moreover, FSA officials could not always detect errors in accountants' and attorneys' statements because the files did not generally contain documentation—such as tax returns—that would have enabled officials to verify the accuracy of the income information in the statements. Although FSA's handbook specifies that files of participants who submit accountants' and attorneys' statements are to contain a minimum set of documentation related to agency procedures and data tracking, these documents do not include income information.

Sometimes, however, accountants' and attorneys' statements included information beyond what is specified in the FSA handbook, and this additional information revealed errors that would not have been evident otherwise. Such additional information included the participant's total income, as well as farm income and nonfarm income; detailed tables with income amounts, sources, and classifications; narrative justifications; or copies of tax returns. For example, we reviewed some statements in which total income amounts revealed calculation errors, or detailed tables revealed that accountants or attorneys had erroneously classified wages or other nonfarm income as farm income. In one file, a statement—which was prepared by a participant himself and explained that the participant was an accountant but not licensed at the time—contained detailed tables revealing that the participant had made errors, such as misclassifying wages as farm income; FSA state officials deemed him eligible to receive 2009 and 2010 farm payments, which he would not have received had the office rejected his statement. In another statement, an accountant included information showing that a participant's income from trading in commodity futures had been classified as farm income—even though the participant had reported this income on his tax returns as unrelated to farming—and this classification brought the participant's nonfarm income below the limit.[24] Relying on the accountant's statement, the FSA state office determined the participant to be eligible to receive over $30,000 in farm payments. In addition, when accountants' or attorneys' statements provided narrative justifications, they sometimes reasoned that certain adjustments should be made that would bring participants' incomes below the limit. For example, one accountant said that FSA should use the income of the participant's deceased mother when verifying eligibility, and another subtracted income that a participant received from an individual retirement account. With such additional explanation, FSA state officials were able to identify the errors, and they did not accept the statements.

In interviews, state officials described taking different approaches to reviewing files with accountants' and attorneys' statements. For example, officials in the Kansas state office said they routinely verified that accountants' and attorneys' license numbers were valid and current, even though such verification was not mentioned in FSA guidance, because they believed it was an important step in holding the accountants and attorneys accountable; officials in several other state offices, in contrast, said they did not do such verification. Also, officials in two state offices said they confirmed that statements included the names of the IRS tax schedules and line numbers where farm income was reported on tax returns but did not further investigate whether it was permissible to report farm income on these lines. When accountants or attorneys enclosed copies of tax returns with their statements, officials in one state office said they reviewed the tax returns, as well as the statements, to verify the accuracy of the statements. In three state offices, in contrast, officials said they reviewed the tax returns only if the statements did not provide enough information to make a determination about a participant's eligibility. Officials in several state offices said they did not feel comfortable challenging anything in an accountant's or attorney's statement because they did not have expertise in accounting or tax preparation and, without additional documentation, had limited support for such a challenge.

FSA Does Not Monitor Its State Offices' Reviews or Verify Accountants' and Attorneys' Statements

FSA's headquarters does not monitor the quality of state offices' reviews of accountants' and attorneys' statements to help ensure that the offices accept only statements that include all the key elements specified in the FSA handbook and are error-free or that the offices retain all required documents in the files, even though, as cited earlier, monitoring is an important federal standard of internal control. Without such monitoring, FSA cannot be assured that eligibility determinations based on state office reviews of accountants' and attorneys' statements are accurate.

To help state officials identify errors in accountants' and attorneys' statements and increase accountability, FSA has made some improvements beginning with reviews for 2011. Specifically, FSA added more detail to its handbook about information that statements are to contain. For example, in addition to the key elements the handbook previously called for, beginning with reviews for 2011, statements were to include the total adjusted gross income used to calculate nonfarm income and documents or a spreadsheet showing how and why joint income was allocated to the relevant individuals. When total adjusted gross income information is available, FSA state officials can identify statements where accountants or attorneys made an error calculating nonfarm income—for example, when an individual's farm income was recorded as less than zero—a mistake one state office reported seeing numerous times. And with a spreadsheet showing how and why joint income was allocated to the relevant individuals, state officials can identify statements where accountants or attorneys made an error in classifying or calculating income amounts. In addition, FSA has developed an automated calculation worksheet to help reduce income calculation errors made during state office reviews. All state offices were to use the new worksheet beginning with reviews for 2011.

Even with these improvements, FSA cannot detect all errors in accountants' and attorneys' statements without reviewing supporting documents. For eligibility requirements other than those based on income (e.g., a requirement that recipients of certain farm payments be "actively engaged in farming"[25]), FSA reviews a sample of program participants' documentation. The agency does not, however perform such reviews to check accountants' and attorneys' statements against documentation supporting participants' eligibility on the basis of income. Consequently, FSA cannot be assured of the statements' accuracy or that payments are being made only to participants whose incomes fall below statutory limits.

FSA IS RECOVERING OVERPAYMENTS, BUT THE NATURAL RESOURCES CONSERVATION SERVICE HAS NOT BEGUN TO DO SO

Since 2012, FSA has been recovering millions of dollars in payments made to participants determined to have exceeded income limits, but NRCS has not started to do so because, according to NRCS officials, the agency first needed to update its project management software. NRCS expects to send letters seeking reimbursement of overpayments by September 2013.

FSA Expects to Recover $143 Million in Payments Made to Participants Who Exceeded Income Limits for 2009 and 2010

According to FSA data, FSA in May 2012 started to recover about $142.7 million in payments made to participants who exceeded income limits for 2009 and 2010, as well as payments to participants who provided no consent forms for 2009 and 2010. As of June 2013, FSA had recovered about $122.2 million in overpayments and sent letters to participants requesting repayment of about $20.5 million.

To recover overpayments made under most FSA programs, agency officials enter ineligibility determinations into the agency's eligibility database; the amount paid to each participant exceeding income limits is then calculated automatically, and letters seeking repayment are sent to these participants. Specifically, FSA uses an integrated payment process, which reads data from many systems to determine whether payments issued to a participant were earned in full or in part. Among other data, these systems hold information on participants' applications for various farm programs, their eligibility status, and their membership in entities such as corporations or partnerships. When a participant's status changes from eligible to ineligible (e.g., because his or her income exceeds limits), the payment process is automatically triggered to recalculate the participant's payment, and if the new payment amount is less than the amount originally paid, an overpayment is identified. After a holding period of about 10 days to provide FSA county offices the opportunity to make corrections if errors are found, the overpayment is automatically transferred to USDA's National Receipts and Receivables System (Receivables System).[26] Once overpayments are in the Receivables System, this system automatically generates debt letters to participants, and FSA officials print and send them from a central location.[27]

For entities with multiple members, each member must meet income limits. By law, when one or more members has an income exceeding the limits, the entity's payment must be reduced in proportion to the ineligible member's ownership share. For such cases, letters are automatically generated and sent to the entity, which must refund the amount attributable to the ineligible member or members.

FSA completed some, but not all, of its income verification reviews for 2009 and 2010 before 2012 farm payments were issued to participants in October 2012. Where possible, the agency used any 2012 farm payments to which a participant was entitled to offset the amount the participant owed. This approach prevented some, but not all, instances in which FSA sent a 2012 payment to a participant in October and then, in the following weeks or months, sent a letter seeking recovery of 2009 or 2010 payments. According to an FSA headquarters official, the agency is limited in its ability to offset payments in this way because it does not receive data from IRS soon enough. FSA and IRS officials told us that they are considering ways for IRS to share data with FSA more often, such as daily. In addition, the timing of some state offices' reviews precluded using offsets for 2012. FSA began implementing the income verification process for 2009 and 2010 in April 2012, when it sent letters to participants notifying them that, according to information reported on tax returns, their incomes may have exceeded limits. Some state offices completed their verification reviews by September or October 2012, but other state offices' reviews took longer, with some reviews still incomplete in January or February 2013. Agency officials in several state offices said it was difficult to complete the reviews more quickly because reviewing tax returns and accountants' or attorneys' statements is time-consuming and requires significant resources.

NRCS Has Not Begun to Recover Overpayments Made to Participants Whose Incomes Exceeded Limits

In contrast to FSA, NRCS has not begun to recover payments made to participants whose incomes exceeded the limits, although agency officials told us that the agency plans to do so soon. Officials said they first had to update the agency's project management software to read data from FSA's eligibility database and automate the payment recovery process, with the goal of making the process accurate, efficient, and properly documented. NRCS headquarters officials said they are in the process of determining the dollar amount of overpayments to be recovered from participants whose incomes exceeded limits for 2009 and 2010 and that the agency expects to know the amount by September 2013.

Unlike FSA's farm payments, which participants must generally apply for each year,[28] NRCS's conservation payments are generally made under multiyear contracts. According to FSA's handbook, participants' compliance with statutory income limits is determined for the year the contract is approved, and once their eligibility is established, participants remain eligible for the term of the contract. Consequently, before collecting overpayments, NRCS must identify any contracts approved in 2009 and 2010 with participants whose incomes exceeded the limit in the year the contract was approved. For all of these contracts—including active, completed, and canceled ones—NRCS must collect all overpayments made in the past and, for the contracts that are still active, prevent payment of any additional amounts. In addition, for contracts that NRCS has with entities, all entity members must meet income limits for the year the contract is approved to be entitled to a payment. Consequently, for contracts approved in 2009 and 2010, NRCS must also identify entities found to have one or more ineligible members for those years and reduce the entity's payment in proportion to the member's share in the entity.

According to NRCS officials, they completed the software update in February 2013 and issued guidance on payment recovery procedures in April 2013. In addition, NRCS held a training session about the new procedures for its state offices, which are responsible for identifying and recovering overpayments made to participants whom FSA determined to have incomes exceeding the conservation limit in 2009 or 2010. The update will allow NRCS's database to coordinate with FSA's eligibility database and automatically prevent future payments to participants who have been recorded as exceeding the income limit in FSA's database. The guidance provides procedures for identifying and collecting overpayments that have already been made. According to this guidance, NRCS's state offices must first update the eligibility status associated with each 2009 and 2010 contract in the agency's contract management database. Next, according to the guidance, state officials are to determine the amounts to be collected for overpayments, send participants collection letters, and enter the receivable amount into the agency's financial management database. The guidance calls for all these steps to be completed by September 2013.

CONCLUSION

As nationwide fiscal pressures continue, and farm incomes remain high, it is crucial to ensure that limited taxpayer dollars are spent to support only eligible farm and conservation program participants. FSA has taken significant steps to ensure that program payments go only to participants whose incomes do not exceed statutory limits. Partly in response to recommendations we made in 2008, the agency implemented management controls whereby (1) it worked with the IRS to identify participants whose incomes potentially exceeded limits laid out in the 2008 Farm Bill and (2) took steps to verify that participants receiving payments had incomes below the limits. FSA now has a process specified in guidance for verifying that participants do not have incomes exceeding limits and, in 2012, started to collect about $143 million in overpayments identified through the new process. Nevertheless, with these provisions of the 2008 Farm Bill still in effect, it remains difficult to accurately classify farm income and nonfarm income when reviewing tax returns. Without simplification of these provisions, this difficulty—and the resulting errors—are likely to persist, along with payments to some ineligible participants. Moreover, FSA state offices have not always followed guidance, and their reviews of tax returns and accountants' or attorneys' statements vary in quality, sometimes resulting in erroneous determinations of participants' eligibility to receive payments. FSA headquarters has not monitored state offices' reviews to assess their quality, even though monitoring is a federal standard for internal control that can help ensure state offices' consistent adherence to guidance, hold state offices accountable for accurate determinations of participants' eligibility to receive payments, and help prevent payments to participants who exceed income limits. Furthermore, accountants' and attorneys' statements sometimes contained errors, which FSA did not consistently identify in the statements we reviewed, in part because the agency did not verify the accuracy of participants' income information for a sample of those statements, as it does for eligibility requirements other than income. Consequently, without monitoring of reviews and verification of statements, errors in FSA's reviews of tax returns and statements are likely to persist, and payments to some ineligible participants are also likely to continue.

MATTER FOR CONGRESSIONAL CONSIDERATION

To reduce the risk of USDA's making potentially improper payments to participants whose incomes exceed statutory limits for farm and conservation programs, Congress should consider simplifying those limits. Such simplification could involve eliminating the distinction between farm income and nonfarm income as reported on tax returns and using total adjusted gross income to set income limits for participants' payment eligibility.

RECOMMENDATIONS FOR EXECUTIVE ACTION

To further improve agency controls that help prevent payments to participants whose incomes exceed eligibility limits, we recommend that the Secretary of Agriculture direct the Administrator of FSA to take the following two actions:

- Institute monitoring to assess the quality of state offices' reviews of tax returns and of accountants' and attorneys' statements to ensure stricter adherence to the agency's guidance for verifying compliance with income limits.
- Implement a process to verify that accountants' and attorneys' statements accurately reflect participants' incomes as reported on income tax returns and supporting documentation or other equivalent documents.

AGENCY COMMENTS AND OUR EVALUATION

We provided the Secretary of Agriculture with a draft of this report for review and comment. In written comments, which are summarized below, the Administrator of FSA said that USDA generally agreed with our report's findings and recommendations. The comment letter further states that agency officials have had discussions under way for development of a methodology to verify the accuracy of information in accountants' and attorneys' statements. USDA also commented that our report does not describe the significant resources— including employees at all levels within FSA—dedicated to the income verification process or the challenges that IRS and FSA worked to overcome to jointly launch this process in 2009. In addition, the letter states that the draft report did not mention the costs to USDA of having IRS use its data, make calculations, and transfer electronic data to FSA. We agree that FSA has devoted significant resources to implementing the verification process—and added information about the costs to USDA of IRS' services—but a detailed review of these resources was beyond the scope of this report.

In addition, USDA stated that our report identifies FSA as responsible for administering the bulk of farm program payments and NRCS as responsible for administering payments for most conservation programs. USDA's letter said, "while this is a true statement, it may mislead the reader to think most of the conservation payments are administered by NRCS, which may not be the case." The comment letter acknowledges that in footnotes the report shows that FSA also administers conservation programs but states that the report "does not distinguish between the administering of programs and payments." When we followed up with NRCS, however, it could not provide data to corroborate this statement or to clarify the number of conservation payments it administers.

USDA stated that it did not agree with the statement in the draft report that NRCS has not begun to recover payments, stating that NRCS officials discussed with us during our review the importance that collections be addressed nationally with a software change to ensure accuracy and consistency. The comment letter states that the payment recovery process began with NRCS issuance of guidance in April 2013. Nevertheless, agency officials said that as of August 2013, NRCS had not yet collected overpayments. Because the agency has not yet collected payments, we continue to believe that our report statement is accurate.

USDA also provided technical comments, which we incorporated as appropriate.

Sincerely yours,

Daniel Garcia-Diaz
Director, Natural Resources and Environment

APPENDIX I: OBJECTIVES, SCOPE, AND METHODOLOGY

Our objectives were to examine (1) the Farm Service Agency's (FSA) review of participants' tax returns to verify that participants receiving payments do not exceed income limits, (2) FSA's review of accountants' and attorneys' statements to verify participants' incomes, and (3) FSA's and the Natural Resources Conservation Service's (NRCS) recovery of overpayments to participants determined to have exceeded income limits.

To examine FSA's review of participants' tax returns and accountants' and attorneys' statements to verify participants' incomes for 2009 and 2010 (the most recent years for which data were available), we reviewed portions of the 2008 Farm Bill related to income limits to understand the statutory requirements, and FSA's handbook on payment eligibility and other relevant agency guidance to understand how the agency implemented the limits. In addition, we reviewed 278 files from a nonprobability sample of 18 FSA state offices chosen from among FSA's 51 state offices. We chose these 18 offices to reflect diversity with respect to number of income eligibility reviews the offices conducted, size and type of farm and conservation programs, and geographic location. We initially selected 19 state offices: 3 of the 4 state offices that had reviewed 750 or more files for income eligibility, 6 of the 35 state offices that had reviewed fewer than 750 files but at least 30, and all 10 of the state offices that had reviewed fewer than 30 files. We later removed 3 of these state offices (Colorado, Georgia, and Minnesota) from our sample when we learned that they had not completed their reviews and, using the same selection criteria above, added 2 others (Indiana and Ohio) that had completed their reviews. The 18 states in our final sample were Arkansas, California, Connecticut, Hawaii, Indiana, Kansas, Louisiana, Maine, Massachusetts, Missouri, Nevada, New Hampshire, New Jersey, Ohio, Rhode Island, Texas, Vermont, and West Virginia.

Under FSA's income verification process, FSA state offices created and reviewed a file for each farm and conservation program participant identified by the Internal Revenue Service (IRS) as potentially exceeding income limits. Because the number of participants identified by IRS varied from state to state, so too did the number of files FSA state offices reviewed, ranging from 0 to nearly 2,000. Using logical break points, we divided the state offices into three groups according to how many files they had reviewed: (1) states with 750 or more files, (2) states with 30 to 749 files, and (3) states with fewer than 30 files. For state offices we selected that had reviewed 750 or more files, we drew a random sample of 30 files for our examination, and for state offices with 30 to 749 files, we generally drew a random sample of 24 files. Within each sample, we used state offices' preliminary data to (1) select half with tax returns and the other half with accountants' and attorneys' statements and (2) include determinations of both eligibility and ineligibility for participation in farm and conservation programs. Once we had sorted state offices' preliminary data into categories on the basis of document type (tax returns or accountants' and attorneys' statements) and eligibility determination, we used computer software to randomly select files from each category.[29] Because we used preliminary data to select our sample, determinations of participants' eligibility sometimes changed, and files originally categorized as tax return files or accountants' or attorneys' statement files sometimes turned out to include both types of documents. For files with both types of documents, we categorized them according to the document used as the basis for FSA's determination of eligibility. For the 10 states with fewer than 30 reviews, we obtained and examined all their files. In total, we reviewed 278 files—115 tax return files and 163 files with accountants' or attorneys' statements. All 18 state offices in our sample had at least 1 file with accountants' or attorneys' statements, and 13 of the state offices had at least 1 tax return file.[30] We reviewed the files to evaluate FSA state offices' compliance with agency guidance and to understand the offices' rationale for their eligibility determinations. Because neither the state offices nor the files we ultimately reviewed were entirely chosen at random, our results cannot be generalized to all files in a given state office (except for the states in which we reviewed all files) or to all files nationwide. Nevertheless, by providing examples, the information from these files enhanced our understanding of FSA's income verification process. To understand the agency's policies, procedures, and practices for verifying participants' income eligibility, and to discuss files we reviewed, we also visited and interviewed FSA headquarters officials and officials in 8 state offices. We selected these state offices to include offices that had at least 250 files to review and to reflect diversity with respect to size and type of farm and conservation program and geographic location. These 8 offices were located in Indiana, Iowa, Kansas, Louisiana, Nebraska, Ohio, Missouri, and Texas.

In addition to reviewing FSA files from selected state offices, we collected nationwide FSA data on the agency's process for determining eligibility on the basis of income. To assess the reliability of FSA's data, we (1) performed electronic testing of required data elements, including examining the data for missing and inconsistent values; (2) reviewed existing information about the data and the system that produced them; and (3) interviewed agency officials knowledgeable about the data. We found these data to be sufficiently reliable for the purposes of this report. We analyzed FSA's data to identify any differences across states in eligibility determination rates for 2010. Specifically, we counted all cases in which FSA made determinations that participants were either eligible or ineligible for farm payments (including when participants acknowledged having incomes that exceeded limits or did not respond to FSA's notification letters). We excluded from this analysis cases in which FSA (1) did not make a determination of eligibility because, for example, the participant did not receive farm payments; (2) removed the participant from review because, for example, he or she was deceased; or (3) determined participants to be ineligible because they did not provide consent forms. We also used the data to report the number of participants identified by IRS as potentially ineligible; the number without consent forms; the number deemed ineligible by FSA; and the dollar amount that FSA expects to collect from participants found to be ineligible for 2009, 2010, and 2011 payments.

To examine FSA's recovery of overpayments to participants whose incomes were determined to have exceeded statutory limits in 2009 and 2010, as well as participants with no consent forms for 2009, 2010, and 2011, we collected nationwide FSA data on payment amounts to be recovered; reviewed agency guidance on payment recovery; and interviewed agency officials about policies, procedures, and practices used to recover such payments. To assess the reliability of these data, we reviewed existing information about the data and the system that produced them and interviewed agency officials knowledgeable about the data. We determined that the data were reliable for purposes of this report. To examine NRCS's payment recovery process, we read government reports on the topic, including reports by the U.S. Department of Agriculture (USDA) Office of Inspector General; reviewed agency guidance; and interviewed NRCS officials in its headquarters and in three state offices (Iowa, Louisiana, and Missouri).[31]

We conducted this performance audit from June 2012 to August 2013 in accordance with generally accepted government auditing standards. Those standards require that we plan and perform the audit to obtain sufficient, appropriate evidence to provide a reasonable basis for our findings and conclusions based on our audit objectives. We believe that the evidence obtained provides a reasonable basis for our findings and conclusions based on our audit objectives.

APPENDIX II: FARM AND CONSERVATION PROGRAM INCOME LIMITS IN THE 2008 FARM BILL

Farm and conservation program participants—individuals, entities, and members of entities—must have incomes below limits established in the 2008 Farm Bill to be eligible to receive payments under certain programs. The 2008 Farm Bill established separate limits for farm income and nonfarm income; both limits are based on adjusted gross income (or a comparable measure) as reported on tax returns and averaged over the 3 most recent years. In addition, the Consolidated and Further Continuing Appropriations Act, 2012, established another income limit applicable only to 2012 recipients of payments known as direct payments.[32] Table 1 shows the programs and the applicable income limit or limits for each, as established in the 2008 Farm Bill and the 2012 act.

The 2008 Farm Bill also outlined the definitions of nonfarm income and farm income. Nonfarm income is the difference between average adjusted gross income and average adjusted gross farm income. Farm income, as defined in USDA regulations implementing the 2008 Farm Bill, is the portion of average adjusted gross income derived from, or related to, the following list of activities:

- production of crops, specialty crops, and unfinished raw forestry products;
- production of livestock, including but not limited to, cattle, elk, reindeer, bison, horses, deer, sheep, goats, swine, poultry, fish and other aquaculture products used for food; honeybees; and products produced by, or derived from, livestock;

Table 1. Income Limits for Farm and Conservation Programs

Program or payment category	Income limit (average adjusted gross income over 3 most recent tax years)		
	Nonfarm income	Farm income	Total income
Farm commodity programs			
Direct and Countercyclical Program countercyclical payments Average Crop Revenue Election Loan Deficiency Payment Marketing Assistance Loan Milk Income Loss Contract	$500,000	N/A	N/A
Direct and Countercyclical Program direct payments, 2009-2012	$500,000	$750,000	N/A
Direct and Countercyclical Program direct payments, 2012	$500,000	$750,000	$1,000,000
Farm disaster programs			
Supplemental Revenue Assistance Noninsured Crop Disaster Assistance Livestock Forage Disaster Livestock Indemnity Emergency Assistance Program for Livestock, Honey Bees and Farm-Raised Fish Tree Assistance	$500,000	N/A	N/A
Conservation programs			
Agricultural Management Assistance Agricultural Water Enhancement Chesapeake Bay Watershed Conservation Reserve Conservation Stewardship Cooperative Conservation Partnership Initiative Environmental Quality Incentives Farm and Ranchland Protection Grasslands Reserve Wetlands Reserve Wildlife Habitat Incentive	$1,000,000[a]	N/A	N/A

Source: GAO analysis of USDA information and legislative provisions.

Legend: N/A = not applicable

Notes: The income limits apply to individuals, legal entities, and members of entities. A husband and wife may divide their income as if they had filed separate income tax returns.

[a] This limit does not apply if at least 66.66 percent of adjusted gross income—that is, the sum of adjusted gross nonfarm income plus adjusted gross farm income—was adjusted gross farm income. FSA or NRCS may waive this limit on a case-by-case basis to protect environmentally sensitive land of special significance.

- production of farm-based renewable energy;
- sale, including the sale of easements and development rights, of farm, ranch, forestry land; water or hunting rights; or environmental benefits;
- rental or lease of land or equipment, used for farming, ranching, or forestry operations, including water or hunting rights;
- processing, packing, storing, shedding, and transporting of farm, ranch, and forestry commodities, including renewable energy;
- feeding, rearing, or finishing of livestock;
- sale of land that has been used for agriculture;
- any payment or benefit, including benefits from risk management practices, crop insurance indemnities, and catastrophic risk protection plans;
- payments and benefits authorized under any program made applicable to income eligibility requirements;
- any other activity related to farming, ranching, or forestry, as determined by FSA;
- any income reported on specified IRS schedules used by the person or legal entity to report income from farming, ranching, or forestry operations to the IRS; and
- sale of equipment to conduct farm, ranch, or forestry operations and the provision of production inputs and services to farmers, ranchers, foresters, and farm operations, if the average adjusted gross farm income is at least 66.66 percent of average adjusted gross income.

APPENDIX III: INFORMATION ON NRCS AND FSA WAIVERS OF STATUTORY INCOME LIMIT FOR CONSERVATION PAYMENTS, 2009 AND 2010

The 2008 Farm Bill allows NRCS or FSA to waive the $1 million nonfarm income limit for conservation payments in cases involving environmentally sensitive land of special significance. According to NRCS documents, the agency waived the income limit 4 times in 2009 and 15 times in 2010, to approve multiyear conservation contracts with payments of about $89 million in total over multiple years. FSA did not issue any waivers in 2009 and waived the limit 11 times in 2010 to approve multiyear contracts with conservation payments of about $80,000 in 2010 and additional payments in subsequent years, according to agency officials. Table 2 shows the number of income limit waivers approved under specified NRCS and FSA programs, and the payment associated with each waiver, for 2009 and 2010.

Table 2. Natural Resources Conservation Service and Farm Service Agency Waivers of Statutory Income Limit and Their Associated Conservation Payments, 2009 and 2010

	2009		2010	
	Number of waivers in year[a]	Benefit value (dollars)	Number of waivers in year[a]	Benefit value (dollars)
NRCS program				
Wetland Reserve Program	2	$10, 000,000	8	$5,800,000
		234,520		21,947,815
				7,924,250
				900,853
				80,000
				31,755,730
				5,405,393
				227,203
Grassland Reserve Program			1	2,797,200
Environmental Quality Incentive Program			4	299,846
				48,768
				210,000
				35,250
Farmland Protection Program	1	$630,000	0	
Wildlife Habitat Incentive Program	1	$193,241	2	99,993
				449,622
Total for NRCS	**4**	**11,057,761**	**15**	**77,981,923**
FSA program				
Conservation Reserve Program	0	0	11	993[b]
				993
				420
				1,945
				6,398
				5,289
				3,277
				9,014
				12,767

	2009		2010	
	Number of waivers in year[a]	Benefit value (dollars)	Number of waivers in year[a]	Benefit value (dollars)
				35,340
				3,448
Total for FSA	**0**	**0**	**11**	**79,884**

Sources: GAO, Farm Service Agency, and Natural Resources Conservation Service.

Notes: Each waiver in this table was granted in the stated year and waived the $1 million nonfarm income limit for a given participant associated with a conservation program contract.

For NRCS, each dollar amount represents the total value of a multiyear contract, which may be paid over many years. For FSA, in contrast, each dollar amount represents only payments made in 2010.

[a] Numbers in this column represent only those waivers associated with contracts and program payments. In 2009, FSA approved three waivers but did not enter into any related contracts, and in 2010, the agency approved one waiver without entering into a related contract.

[b] For FSA's Conservation Reserve Program, dollar amounts represent only the portion of each contract for which a payment was made in 2010. According to an FSA official, total values of multiyear contracts were not readily available.

APPENDIX IV: NUMBER OF FARM AND CONSERVATION PROGRAM PARTICIPANTS INELIGIBLE BECAUSE OF INCOME LIMITS, 2009 AND 2010

In 2009, FSA implemented a new process for verifying whether farm and conservation program participants' incomes exceeded statutory limits and determined that less than 1 percent of participants receiving payments under these programs were in fact ineligible for the payments. Under the new process, participants provided consent forms allowing IRS to disclose certain tax-related information to FSA.

For participants who provided the consent forms, IRS identified and shared with FSA a list of participants for whom tax data showed that their incomes may have exceeded statutory limits in 2009, 2010, or both years. FSA also identified additional participants who had no consent forms in those years. According to FSA's handbook, participants who do not provide consent forms are ineligible to receive farm and conservation payments through programs subject to income limits.

FSA gave all participants identified by IRS or FSA an opportunity to provide additional information and used any information provided to determine eligibility. For participants determined to be ineligible for 2009 and 2010 FSA program payments, the agency identified overpayment amounts and sent letters seeking repayment. For 2009 and 2010, tables 3 and 4, respectively, show, by state, the number of program participants and the number of participants IRS identified as potentially exceeding income limits.

Table 3. Farm and Conservation Program Participants Found to Be Ineligible in 2009, by State

State	Number of participants subject to income limits	Number of participants identified by IRS as potentially exceeding income limits	Number of participants without consent forms	Number of participants ineligible because of income limits
Alabama	19,631	596	2,298	441
Alaska	60		19	1
American Samoa			4	
Arizona	2,203	309	1,129	102
Arkansas	26,290	767	2,311	459
California	13,396	1,264	4,585	307
Colorado	19,359	519	2,474	159
Connecticut	330	15	163	52
Delaware	1,221	55	218	21
Florida	4,722	343	1,490	220
Georgia	26,551	689	2,906	197
Guam		.	6	1
Hawaii	194	24	175	28
Idaho	12,972	416	1,157	161
Illinois	126,878	1,997	7,160	1,080
Indiana	64,822	844	4,017	512
Iowa	118,593	1,315	6,176	692
Kansas	102,365	1,677	6,104	1,068
Kentucky	63,403	679	1,469	708
Louisiana	24,250	623	2,553	1,071
Maine	1,462	12	1,336	241
Maryland	6,003	229	399	94
Massachusetts	547	25	250	18

State	Number of participants subject to income limits	Number of participants identified by IRS as potentially exceeding income limits	Number of participants without consent forms	Number of participants ineligible because of income limits
Michigan	28,806	191	2,148	231
Minnesota	75,969	569	3,949	382
Mississippi	25,428	672	3,347	229
Missouri	69,063	1,097	4,800	826
Montana	20,241	407	2,209	145
Nebraska	67,727	918	1,676	338
Nevada	337	28	154	17
New Hampshire	279	20	343	75
New Jersey	910	46	93	19
New Mexico	5,033	165	1,682	471
New York	11,488	95	1,481	120
North Carolina	30,746	413	1,918	173
North Dakota	42,312	351	1,945	151
Northern Mariana			5	1
Ohio	57,425	501	3,162	208
Oklahoma	40,425	803	4,974	1,248
Oregon	7,202	278	1,336	81
Pennsylvania	17,626	108	1,466	63
Puerto Rico	635		938	
Rhode Island	49	10	171	
South Carolina	12,062	286	1,104	297
South Dakota	39,173	597	2,200	297
Tennessee	36,546	543	2,428	394
Texas	93,466	3,021	8,796	2,353
Utah	3,260	76	678	131
Vermont	1,746	26	683	65
Virgin Islands	17			.
Virginia	15,681	317	404	107
Washington	14,130	337	2,442	160
West Virginia	2,300	36	220	30
Wisconsin	51,661	314	4,287	293
Wyoming	3,514	154	701	66
Total	**1,410,647**	**24,777**	**110,148**	**16,604**

Source: GAO analysis of FSA data.

Note: Numbers of participants include individuals, legal entities, and members of entities, all of whom are subject to income limits under the 2008 Farm Bill.

Table 4. Farm and Conservation Program Participants Found to Be Ineligible in 2010, by State

State	Number of participants subject to income limits	Number of participants identified by IRS as potentially exceeding income limits	Number of participants without consent forms	Number of participants ineligible because of income limits
Alabama	18,785	1,986	351	591
Alaska	58	196	8	.
American Samoa		7		.
Arizona	3,113	287	884	53
Arkansas	26,191	775	1,808	389
California	12,970	1,228	3,641	246
Colorado	19,548	523	2,079	108
Connecticut	380	14	135	39
Delaware	1,242	57	147	9
Florida	4,999	361	1,499	169
Georgia	24,863	713	2,444	153
Guam	1		4	1
Hawaii	348	27	210	24
Idaho	12,740	419	854	128
Illinois	128,527	2,003	5,263	851
Indiana	64,163	853	2,938	380
Iowa	118,362	1,366	4,753	490
Kansas	102,006	1,734	4,759	776
Kentucky	63,217	631	1,064	523
Louisiana	22,922	635	2,069	693
Maine	1,512	25	664	254
Maryland	6,029	223	295	79
Massachusetts	598	38	211	13
Michigan	28,852	217	1,632	158
Minnesota	75,656	582	3,154	287
Mississippi	25,123	728	2,812	194
Missouri	68,662	1,132	3,622	549
Montana	20,558	408	1,564	88
Nebraska	67,492	940	1,238	260
Nevada	343	27	116	11
New Hampshire	364	19	287	97
New Jersey	1,004	56	100	18
New Mexico	5,730	183	1,460	431

State	Number of participants subject to income limits	Number of participants identified by IRS as potentially exceeding income limits	Number of participants without consent forms	Number of participants ineligible because of income limits
New York	11,579	113	1,251	103
North Carolina	29,168	413	1,317	137
North Dakota	42,982	377	1,428	116
Northern Mariana			11	2
Ohio	56,931	522	2,318	168
Oklahoma	39,978	818	3,982	1,121
Oregon	7,467	293	1,197	107
Pennsylvania	17,884	107	1,319	80
Puerto Rico	716		1,351	
Rhode Island	55	12	157	
South Carolina	11,276	296	869	208
South Dakota	39,156	594	1,616	232
Tennessee	33,316	538	1,918	314
Texas	91,591	3,084	6,501	1,768
Utah	3,311	98	534	98
Vermont	1,730	41	720	84
Virgin Islands	3		30	
Virginia	15,797	299	382	103
Washington	14,483	350	2,113	134
West Virginia	2,611	40	252	17
Wisconsin	50,825	327	3,321	204
Wyoming	3,516	160	531	52
Total	**1,400,860**	**25,277**	**87,013**	**12,878**

Source: GAO analysis of FSA data.

APPENDIX V: FARM PROGRAM PARTICIPANTS DETERMINED TO BE ELIGIBLE FOR DIRECT PAYMENTS IN 2009 AND 2010

For 2009 and 2010, FSA's state offices reviewed documentation for farm program participants identified by IRS as having incomes that may have exceeded limits for farm income or nonfarm income. Through these reviews, state offices determined whether participants were eligible for payments under various farm programs, including eligibility for payments known as direct payments. These payments consist of fixed annual sums based upon a farm's historical acreage planted in particular commodity crops specified in legislation. We chose to include information on direct payments because during 2009 and 2010, these payments accounted for the largest dollar amounts paid by FSA to participants in its farm programs. For 2009 and 2010, respectively, tables 5 and 6 show the number of participants reviewed by each state office and the percentage determined to be eligible.

Table 5. Participants Determined to Be Eligible for Direct Payments in 2009, by State

State	Number of participants reviewed	Number of eligible participants	Percentage eligible	Number of ineligible participants	Percentage ineligible
Alabama	228	93	40.8%	135	59.2%
Arizona	173	138	79.8	35	20.2
Arkansas	353	219	62.0	134	38.0
California	565	420	74.3	145	25.7
Colorado	49	48	98.0	1	2.0
Connecticut	3	1	33.3	2	66.7
Delaware	23	11	47.8	12	52.2
Florida	59	39	66.1	20	33.9
Georgia	201	114	56.7	87	43.3
Idaho	185	120	64.9	65	35.1
Illinois	801	327	40.8	474	59.2
Indiana	282	152	53.9	130	46.1
Iowa	413	228	55.2	185	44.8
Kansas	598	282	47.2	316	52.8
Kentucky	249	96	38.6	153	61.4
Louisiana	257	102	39.7	155	60.3
Maryland	87	34	39.1	53	60.9
Massachusetts	3	1	33.3	2	66.7
Michigan	92	38	41.3	54	58.7
Minnesota	198	121	61.1	77	38.9
Mississippi	259	165	63.7	94	36.3
Missouri	464	176	37.9	288	62.1
Montana	68	46	67.6	22	32.4
Nebraska	334	156	46.7	178	53.3

State	Number of participants reviewed	Number of eligible participants	Percentage eligible	Number of ineligible participants	Percentage ineligible
Nevada	7	2	28.6	5	71.4
New Hampshire	2	2	100.0		0.0
New Jersey	12	10	83.3	2	16.7
New Mexico	32	21	65.6	11	34.4
New York	34	22	64.7	12	35.3
North Carolina	156	69	44.2	87	55.8
North Dakota	113	54	47.8	59	52.2
Ohio	130	23	17.7	107	82.3
Oklahoma	302	132	43.7	170	56.3
Oregon	127	78	61.4	49	38.6
Pennsylvania	35	18	51.4	17	48.6
South Carolina	107	35	32.7	72	67.3
South Dakota	209	96	45.9	113	54.1
Tennessee	191	82	42.9	109	57.1
Texas	1,186	569	48.0	617	52.0
Utah	38	19	50.0	19	50.0
Vermont	10	7	70.0	3	30.0
Virginia	67	41	61.2	26	38.8
Washington	90	72	80.0	18	20.0
West Virginia	16	6	37.5	10	62.5
Wisconsin	117	55	47.0	62	53.0
Wyoming	40	34	85.0	6	15.0
Total	**8,965**	**4,574**	**51.0%**	**4,391**	**49.0%**

Source: GAO analysis of FSA data.

Table 6. Participants Determined to Be Eligible for Direct Payments in 2010, by State

State	Number of participants reviewed	Number of eligible participants	Percentage eligible	Number of ineligible participants	Percentage ineligible
Alabama	211	82	38.9	129	61.1
Arizona	135	107	79.3	28	20.7
Arkansas	377	229	60.7	148	39.3

Table 6. (Continued)

State	Number of participants reviewed	Number of eligible participants	Percentage eligible	Number of ineligible participants	Percentage ineligible
California	522	388	74.3	134	25.7
Colorado	66	66	100.0	.	0.0
Connecticut	6	4	66.7	2	33.3
Delaware	16	10	62.5	6	37.5
Florida	56	37	66.1	19	33.9
Georgia	184	112	60.9	72	39.1
Hawaii	3	2	66.7	1	33.3
Idaho	179	125	69.8	54	30.2
Illinois	787	326	41.4	461	58.6
Indiana	309	175	56.6	134	43.4
Iowa	446	253	56.7	193	43.3
Kansas	605	302	49.9	303	50.1
Kentucky	230	101	43.9	129	56.1
Louisiana	275	135	49.1	140	50.9
Maryland	78	37	47.4	41	52.6
Massachusetts	7	.		7	.
Michigan	98	37	37.8	61	62.2
Minnesota	247	165	66.8	82	33.2
Mississippi	241	166	68.9	75	31.1
Missouri	451	198	43.9	253	56.1
Montana	68	49	72.1	19	27.9
Nebraska	327	162	49.5	165	50.5
Nevada	5	1	20.0	4	80.0
New Hampshire	2	2	100.0		0.0
New Jersey	15	10	66.7	5	33.3
New Mexico	28	19	67.9	9	32.1
New York	40	26	65.0	14	35.0
North Carolina	150	78	52.0	72	48.0
North Dakota	141	78	55.3	63	44.7
Ohio	152	40	26.3	112	73.7
Oklahoma	344	147	42.7	197	57.3
Oregon	131	81	61.8	50	38.2
Pennsylvania	45	29	64.4	16	35.6
South Carolina	97	35	36.1	62	63.9
South Dakota	222	121	54.5	101	45.5

State	Number of participants reviewed	Number of eligible participants	Percentage eligible	Number of ineligible participants	Percentage ineligible
Tennessee	205	85	41.5	120	58.5
Texas	1,240	619	49.9	621	50.1
Utah	29	14	48.3	15	51.7
Vermont	8	5	62.5	3	37.5
Virginia	57	26	45.6	31	54.4
Washington	96	80	83.3	16	16.7
West Virginia	10	3	30.0	7	70.0
Wisconsin	111	54	48.6	57	51.4
Wyoming	34	30	88.2	4	11.8
Total	**9,086**	**4,851**	**53.4%**	**4,235**	**46.6%**

Source: GAO analysis of FSA data.

End Notes

[1] Throughout this report, we use the term *participants* to refer to individuals as well as entities, such as corporations, estates, and trusts, that receive farm and conservation payments.

[2] FSA is responsible for administering the Conservation Reserve Program and other conservation programs as well.

[3] Specifically, participants are ineligible for payments known as direct payments if their farm incomes exceed $750,000. Direct payments are fixed annual sums based on a farm's historical production of particular commodity crops.

[4] Pub. L. No. 110-246 § 1604(a), 110 Stat. 1651, 1741 (amending 7 U.S.C. § 1308-3a(b)).

[5] GAO, *Federal Farm Programs: USDA Needs to Strengthen Controls to Prevent Payments to Individuals Who Exceed Income Eligibility Limits*, GAO-09-67 (Washington, D.C.: Oct. 24, 2008).

[6] FSA determines for both agencies whether participants' incomes exceed limits, but FSA does not make determinations of participants' eligibility for NRCS programs.

[7] Participants may choose to send either (1) tax returns or (2) accountants' or attorneys' statements to support their eligibility for farm and conservation payments. As a result, a given FSA state office does not necessarily review both types of files. Thirteen of the 18 FSA state offices in our sample had reviewed both types of files, and all 18 had reviewed files containing accountants' and attorneys' statements. The 13 state offices that had reviewed both types of files were in Arkansas, California, Connecticut, Indiana, Kansas, Louisiana, Missouri, Nevada, New Jersey, Ohio, Texas, Vermont, and West Virginia. The 5 state offices that had reviewed files containing only accountants' or attorneys' statements were in Hawaii, Maine, Massachusetts, New Hampshire, and Rhode Island.

[8] Adjusted gross income is defined as taxable income from all sources, including earned income, such as wages and salaries, and unearned income, such as interest or dividends, minus specific deductions.

[9] Pub. L. No. 110-246 § 1604(a), 110 Stat.1651, 1742 (amending 7 U.S.C. § 1308-3a(b)(2)(A)(ii)).

[10] In 2009, FSA approved three waivers but did not enter into any related contracts. The waivers FSA approved in 2010 were for multiyear contracts, but the agency could not readily provide us the total, multiyear value of these contracts.

[11] GAO

[12] IRS does not provide to FSA dollar amounts of income reported on participants' tax returns, average income amounts, or determinations of participants' eligibility. Information that FSA receives from IRS is stored in a secure database, only accessible to authorized personnel, used only in the income verification process, and not subject to release under the Freedom of Information Act.

[13] For two reasons, IRS's lists of participants whose incomes may exceed 2008 Farm Bill income limits may also include participants whose incomes do not, in fact, exceed those limits: (1) IRS does not allocate income to individual participants from taxpayer returns when couples file joint returns, and (2) IRS does not always classify farm income and nonfarm income the same way that FSA does to determine participants' eligibility for payments because the two agencies' definitions of farm income and nonfarm income differ.

[14] USDA, Farm Service Agency, *FSA Handbook: Payment Eligibility, Payment Limitation, and Average Adjusted Gross Income*, 4-PL, Amendment 16 (Washington, D.C.: 2012).

[15] USDA, Office of Inspector General, *Natural Resources Conservation Service's Oversight and Compliance Activities*, 10601-0001-22 (Washington, D.C.: Feb. 7, 2013).

[16] USDA reimburses IRS annually for services related to the income verification process, including accessing tax information, executing required calculations and comparisons, and transferring electronic data to FSA. In August 2013, USDA reported that since 2009, it had paid IRS $8.69 million for these services.

[17] Beginning with 2012, FSA used a single form for participants to certify their compliance with income limits and consent for IRS to disclose certain tax-related information to FSA. The certification form must be on file for a participant to receive any payments, so using only one form for both purposes helps ensure that participants do not receive payments without submitting a consent form.

[18] About 1,800 of the 15,500 participants identified by IRS were ineligible for conservation payments because their nonfarm incomes exceeded $1 million.

[19] According to the Congressional Research Service, the Senate bill would tighten the income limit for most individuals, but it could restore program eligibility for some individuals with nonfarm incomes from $500,000 to $750,000 if their farm incomes are low. The House bill would tighten the income limit for some individuals, and for other individuals, it could restore program eligibility.

[20] USDA administers the federal crop insurance program in partnership with private insurance companies. Under the program, participants can insure against losses caused by poor crop yields, declines in crop prices, or both for each insurable crop they produce. In 2012, the program provided about $117 billion in insurance coverage for over 1 million policies. Program costs include subsidies to pay for part of participants' premiums.

[21] GAO, *Standards for Internal Control in the Federal Government*, GAO/AIMD-00-21.3.1 (Washington, D.C.: November 1999).

[22] According to FSA officials, pass-through farm income that a participant receives from an S-corporation, partnership, or limited liability company is to be considered farm income.

[23] According to FSA officials, they worked with the American Institute of Certified Public Accountants to develop the handbook's instructions for accountants' and attorneys' statements. This organization represents the accounting profession and sets certain professional standards.

[24] In this case, the relevant provisions of the IRS tax code and FSA guidance are subject to interpretation, so the accountant's classification of this participant's income, as well as FSA's eligibility determination, may have been proper. According to IRS's *Farmer's Tax Guide*, a farmer is to count income from trading in commodity futures as farm income if the following three conditions are met: (1) the farmer enters the transactions to protect himself

from the risk of unfavorable price fluctuations, (2) the commodity is the same one the farmer produces, and (3) the amount traded is within the farmer's range of production. According to the *Farmer's Tax Guide*, such income is to be reported as farm income on tax returns. In general, when income is generated through trading in commodity futures unrelated to a farmer's own commodity production, the *Farmer's Tax Guide* calls for such income to be reported as capital gains, which are taxed at a lower rate.

[25] To be eligible for payments and benefits under certain farm programs, all program participants, either individuals or legal entities, must provide significant contributions to the farming operation to be considered actively engaged in farming. Contributions can consist of capital, land, equipment, or all three, as well as active personal labor, active personal management, or both. Each year, FSA requests documentation from a sample of participants to verify that they meet this "actively engaged" requirement. For example, to verify labor contributions, FSA may request employee time sheets or books or canceled checks paid to hired labor.

[26] FSA has about 2,100 county offices, which are responsible for recording participant information in the agency's eligibility database and for taking certain steps related to collecting overpayments, among other responsibilities.

[27] For payments under programs such as certain small conservation and loan programs, FSA county officials manually calculate and enter into the Receivables System the amounts to be recovered from participants whose incomes exceed the limits, and letters are automatically generated.

[28] For the Conservation Reserve Program, which has multiyear contracts and is one of the conservation programs FSA administers, FSA is responsible for recovering payments from participants with incomes exceeding $1 million who had contracts approved under this program in 2009 or 2010.

[29] For one state office, we drew the random sample by hand, selecting files at specific intervals from a list of all files in each category, because we had not yet developed the approach using computer software.

[30] The 13 state offices that had at least one tax return file were in Arkansas, California, Connecticut, Indiana, Kansas, Louisiana, Missouri, Nevada, New Jersey, Ohio, Texas, Vermont, and West Virginia.

[31] We selected these three NRCS state offices because of their proximity to FSA state offices we visited.

[32] Direct payments are fixed annual payments based on farms' historical production of particular commodity crops.

In: Farm Programs
Editor: Edward A. Roeder
ISBN: 978-1-62948-622-2

Chapter 2

FARM PROGRAMS: USDA NEEDS TO DO MORE TO PREVENT IMPROPER PAYMENTS TO DECEASED INDIVIDUALS*

United States Government Accountability Office

WHY GAO DID THIS STUDY

USDA spends about $20 billion annually on federal programs that support farm income, conserve natural resources, and help farmers manage risks from natural disasters, benefiting over 1 million participants. Given their cost and continuing nationwide budget pressures, these programs have come under increasing scrutiny. One concern has been the distribution of benefits to ineligible participants, including potentially improper payments to deceased individuals, which, as GAO and others have reported, may call into question whether these farm safety net programs are benefiting the agricultural sector as intended. GAO was asked to evaluate USDA controls over payments to the deceased. This report examines the extent to which procedures are in place to prevent (1) FSA and (2) NRCS from making potentially improper payments to deceased individuals and (3) RMA from providing potentially improper subsidies on behalf of deceased individuals. GAO reviewed a random sample

* This is an edited, reformatted and augmented version of the Highlights of GAO-13-503, a report to the Chairwoman, Committee on Agriculture, Nutrition and Forestry, U.S. Senate, dated June 2013.

of payments, compared USDA's databases with SSA's master list of deceased individuals, and interviewed agency officials.

WHAT GAO RECOMMENDS

GAO recommends that FSA further strengthen its verification of payments to deceased individuals, NRCS develop and implement procedures to prevent improper payments to deceased individuals, and RMA develop and implement procedures to prevent improper crop insurance subsidies on behalf of deceased policyholders and to improve the effectiveness of its data mining. USDA generally agreed with GAO's findings and recommendations.

WHAT GAO FOUND

Since 2007, the Department of Agriculture's (USDA) Farm Service Agency (FSA), which administers various programs for farmers that help support farm incomes and provide disaster assistance, has established procedures for preventing improper payments to deceased individuals, including, on a quarterly basis, matching payments to program participants with the Social Security Administration's (SSA) data on deceased individuals. In addition, FSA state and county offices review and verify whether payments made to deceased individuals are proper or improper. Overall, these procedures have enabled FSA to identify thousands of deceased individuals who were paid $3.3 million in improper payments after their dates of death, of which FSA has recovered approximately $1 million. GAO reviewed a generalizable random sample of payments to deceased individuals that FSA identified as proper and found that 9 percent did not have sufficient support to be coded as proper. More monitoring to ensure that county offices' coding of payments is supported by documentation could help reduce the error rate.

The Natural Resources Conservation Service (NRCS), a USDA agency that administers voluntary conservation programs, does not have procedures to prevent potentially improper payments to deceased individuals. For example, NRCS's ability to verify whether payment recipients have died is limited because the agency does not match these recipients against SSA's master list of deceased individuals. Under the standards for internal control in the federal government, agencies are to clearly document such control in the form of management directives, administrative policies, or operating manuals. GAO did a data review for fiscal year 2008 to April 2012, and estimates that NRCS made $10.6 million payments on behalf of 1,103 deceased individuals 1 year or more after their death. Some of these payments may have been proper, but NRCS cannot be certain because it neither identifies which of its payments were made to deceased individuals, nor reviews each of these payments.

USDA's Risk Management Agency (RMA), which administers crop insurance programs, does not have procedures in place consistent with federal internal control standards to prevent potentially improper subsidies on behalf of deceased individuals. For example, RMA does not use SSA's master list of deceased individuals to verify whether its policyholders have died. GAO matched every policyholder's Social Security number in RMA's crop insurance subsidy and administrative allowance data for crop insurance years 2008 to 2012 with SSA's master list of deceased individuals and found that $22 million in subsidies and allowances may have been provided on behalf of an estimated 3,434 program policyholders 2 or more years after death. Many of these subsidies and allowances may have been proper, but without reviewing each subsidy and allowance made on behalf of deceased individuals, RMA cannot be certain that these subsidies and allowances are proper. In addition, without accurate records of which policyholders are deceased, RMA may be less likely to rely on results from data mining—a technique for extracting knowledge from large volumes of data—and therefore be less likely to detect fraudulent, wasteful, or abusive crop insurance claims.

ABBREVIATIONS

FSA	Farm Service Agency
NRCS	Natural Resources Conservation Service
RMA	Risk Management Agency
SSA	Social Security Administration
USDA	U.S. Department of Agriculture

June 28, 2013

The Honorable Debbie Stabenow
Chairwoman
Committee on Agriculture, Nutrition and Forestry
United States Senate

Dear Madam Chairwoman:

The U.S. Department of Agriculture (USDA) spends about $20 billion annually on federal programs that support farm income, conserve natural resources, and help farmers and farming entities manage risks from natural disasters. These programs benefit over 1 million participants. Given their cost to the federal government, these agricultural safety net programs have come under increasing scrutiny over the years, particularly in light of the nation's severe budget pressures and high farm income levels. One area of concern has been the distribution of benefits to ineligible participants, including potentially improper payments to deceased individuals, which, as we and others have reported, may call into question whether these safety net programs are benefiting the agricultural sector as intended.

Three USDA agencies—the Farm Service Agency (FSA), the Natural Resources Conservation Service (NRCS), and the Risk Management Agency (RMA)—play key roles in aiding farmers. FSA administers various programs that help support farm income and provide assistance after disasters. NRCS works with farmers and other landowners to implement practices to conserve natural resources, such as soil and water. RMA administers the federal crop insurance program, which subsidizes crop insurance to farmers to mitigate losses due to poor crop yields, declines in crop prices, or both. Both FSA and NRCS provide payments to farmers and farming entities through a network of agency headquarters, state, and county offices. RMA, in contrast, administers crop insurance to farmers through a network of 17 approved insurance companies. In 2012, FSA provided $7.4 billion for farm commodity and conservation programs, NRCS provided $4.0 billion for financial and technical assistance in conservation efforts, and RMA provided $8.3 billion in subsidies and allowances for crop insurance.

Over the years, we and USDA's Office of Inspector General have reported on the risks of fraud, waste, and abuse in these programs and recommended ways to alleviate such risks.[1] Specifically, a 2013 study by USDA's Office of Inspector General found that FSA, NRCS, and RMA were among seven USDA agencies with high-risk programs vulnerable to significant improper payments. In a July 2007 report, we found that from 1999 through 2005, FSA paid $1.1 billion in farm payments in the names of 172,801 deceased individuals, and we made recommendations for improving FSA's ability to prevent improper payments.[2] Among other things, we recommended that FSA implement management controls to verify that individuals receiving payments have not died and to determine if improper payments were made to deceased individuals or to entities, such as corporations or partnerships, that did not disclose the death of a member and, if so, to recover the appropriate amounts. Subsequently, the Food, Conservation, and Energy Act of 2008 (2008 Farm Bill) was enacted, requiring FSA to reconcile records of individuals who received payments with Social Security Administration (SSA) data twice each year to determine if the individuals are alive. In addition, we also reported in April 2012 on weaknesses in RMA's crop insurance program, and we screened our past work to identify principles significant to integrity, effectiveness, and efficiency in farm bill programs.[3] One of these principles, framed as a question, applies to oversight of these programs by federal agencies: "Does the program have mechanisms, such as management controls, to monitor compliance and help minimize fraud, waste, and abuse in areas where these are most likely to occur?" In an April 16, 2013, hearing before the House Committee on Appropriations, Subcommittee on Agriculture, Rural Development, Food and Drug Administration, and Related Agencies, the Secretary of Agriculture commented that a reduction in crop insurance errors, including fraud, could save tens of millions of dollars. Further, the Improper Payments Information Act of 2002, as amended, requires federal agencies to periodically review their programs and to identify those that may be susceptible to significant improper payments, including payments and insurance subsidies to ineligible individuals.[4] In addition, the Improper Payments Elimination and Recovery and Improvement Act of 2012 gives specific statutory authority to the Do Not Pay initiative, requiring federal agencies to review prepayment procedures and available databases with relevant information on eligibility, including SSA's death master file, to determine program eligibility and prevent improper payments.[5] Federal law and USDA regulations also require that certain participants be actively engaged in farming to qualify for farm program payments.

In this context, you asked us to evaluate USDA's guidance and control over payments to deceased individuals and the appropriateness of such payments. This report examines the extent to which procedures are in place to prevent (1) FSA from making potentially improper payments to deceased individuals, (2) NRCS from making potentially improper payments to deceased individuals, and (3) RMA from providing potentially improper subsidies on behalf of deceased individuals.

To address all three objectives, we reviewed the 2008 Farm Bill, USDA regulations and guidance, and relevant reports by us and USDA's Inspector General; we also interviewed officials about what steps the agencies take to prevent potentially improper payments or benefits to deceased individuals. We interviewed USDA officials in Washington, D.C.; California; Illinois; Kansas; Missouri; and Texas to obtain information about agency procedures and adherence to guidance. We selected these locations to reflect a variety of characteristics, such as a high number of FSA payments made to deceased individuals, and geographic diversity. To address the first objective, we also interviewed FSA officials about the extent to which this agency follows requirements to compare agency data with the data in SSA's death master file and the steps the agency takes to recover improper payments.[6] To determine if state and county offices properly coded payments made to deceased individuals (coding designates a given payment as proper or improper), we used a generalizable, random sample of 100 payments made by FSA to deceased individuals for the most recent four consecutive quarters of data available—starting in the third quarter of fiscal year 2011 (i.e., July 2011)—after FSA implemented a web-based system for reviewing payments to deceased individuals. We also reviewed and analyzed supporting hard-copy documentation to determine the individuals' eligibility. To understand the reasons for potential errors, we generated a nonprobability sample of 20 payments made to deceased individuals for the same period and deemed erroneous by the agency, and we reviewed and analyzed the supporting documentation for these improper payments as well. Because this sample was a nonprobability sample, the results cannot be generalized to all recipients of FSA payments, but they offer examples of improper payments to deceased individuals. To address the second objective, we compared NRCS payment data since the 2008 Farm Bill—from fiscal year 2008 through April 2012—with the death master file to determine the extent to which NRCS has made potentially improper payments to deceased individuals. To address the third objective, we interviewed RMA agency officials to obtain information about the steps the agency takes to verify its records of eligible policyholders. To determine the extent to which RMA ensures that its records are regularly updated to verify whether policyholders have died, we compared the death master file with crop insurance data from RMA for crop reinsurance year 2008 through reinsurance year 2012.[7] For all three agencies' payment information and for SSA's death master file, we found that their data were sufficiently reliable for the purposes of our review. Appendix I presents a more detailed discussion of our scope and methodology.

We conducted this performance audit from March 2012 through June 2013 in accordance with generally accepted government auditing standards. Those standards require that we plan and perform the audit to obtain sufficient, appropriate evidence to provide a reasonable basis for our findings and conclusions based on our audit objectives. We believe that the evidence obtained provides a reasonable basis for our findings and conclusions based on our audit objectives.

BACKGROUND

USDA provides various programs that assist farmers and landowners through its subsidiary agencies FSA, NRCS, and RMA.

FSA Provides Benefits through Commodity and Crop Disaster Assistance Programs

FSA provides benefits to farmers through various programs, including farm commodity and crop disaster assistance programs authorized in the 2008 Farm Bill. FSA has overall responsibility for administering these programs, including ensuring that all recipients meet eligibility requirements and do not receive payments in excess of program limitations. Farming operations—whether individuals or entities—applying for benefits must file a farm operating plan and an annual acreage report with their local field office, and if any changes occur that could affect program eligibility, such as changes affecting one's status as actively engaged in farming, the farming operation must file a revised farm operating plan. These documents record farming information, such as the name of each individual with an interest in the farming operation, which crops are planted on each field, and the farming practices used. FSA uses this information to determine farm program payments, including payments for various agricultural disaster assistance programs. Most farmers receive farm program payments directly from FSA as individual operators. Some farmers, however, use legal entities to organize their farming operations, thereby reducing their exposure to financial liabilities. Many FSA farm programs have statutory payment limits that set the maximum payment amount that an individual can receive per year. For example, for one of FSA's farm commodity programs, the annual payment limit is $40,000 per individual. FSA carries out these responsibilities through its headquarters office, 50 state offices, and over 2,100 field offices.

In a 2007 report, we recommended that FSA implement management controls, such as matching payment files with SSA's death master file, to verify that an individual receiving payments has not died and to provide reasonable assurance that the agency does not make improper payments to deceased individuals. We also recommended that FSA determine if improper program payments have been made to deceased individuals or to entities that failed to disclose the death of a member and, if so, recover the appropriate amounts.[8]

The 2008 Farm Bill required FSA (1) to promulgate regulations that (a) describe the circumstances under which payments may be issued in the name of a deceased individual and (b) preclude the issuance of payments to, and on behalf of, deceased individuals who were not eligible for the payments and (2) to at least twice each year reconcile SSA's numbers of all individuals who receive commodity program payments.

NRCS Administers Conservation Programs

NRCS administers voluntary programs that offer financial and technical assistance to eligible landowners and producers to help manage natural resources in a sustainable manner. Through these programs, the agency approves conservation program contracts or conservation easements, which provide participants financial assistance for planning and implementing conservation practices to save energy and improve resources, such as soil and water on agricultural lands and on nonindustrial private forestland. These programs can be divided into two categories—working-lands programs and easement programs.[9] Under working-lands programs, NRCS provides program participants with financial and technical assistance through contracts, which are generally in effect for up to 10 years, to implement conservation practices on agricultural land and nonindustrial private forestland. Under easement programs, NRCS purchases conservation easements to restore, protect, or enhance grasslands and wetlands and provides assistance to eligible entities to purchase development rights to keep productive farm and ranch lands in agricultural uses. According to NRCS officials, when an NRCS program participant dies, NRCS considers the contract canceled, unless the participant had identified an executor or other estate representative to act on his or her behalf to transfer the contract to an eligible successor or to complete the contracted activities.

RMA Administers Subsidized Crop Insurance

By promoting crop insurance, the federal government has played an active role in helping to mitigate the effects on income of production risks—droughts, floods, and other natural disasters—as well as price risks that farmers face.

RMA administers the federal crop insurance program, including controlling costs and protecting against fraud, waste, and abuse. The agency partners with 17 approved insurance companies, which sell and service the federal program's insurance policies and share in a percentage of the associated risk of loss and opportunity for gain. The federal government subsidizes about 60 percent of the insurance premiums the insurance companies charge farmers and also pays the companies an allowancc for administrative expenses, which is intended to cover the companies' expenses for selling crop insurance policies and providing customer service to policyholders.

Unlike payments made by FSA and NRCS, subsidies for crop insurance premiums are not paid directly to policyholders but can be considered a financial benefit to them. Without a premium subsidy, a policyholder would have to pay the full amount of the premium. The allowances for administrative expenses can be considered a further benefit to these policyholders, since these allowances are paid on their behalf. RMA also shares with insurance companies payments made on policyholders' claims for losses.

FSA Has Established Procedures for Preventing Improper Payments to the Deceased, but Some Supporting Information Is Incomplete

Since 2007, FSA has established procedures for preventing improper payments to deceased individuals, including matching payments to program participants with SSA's data on deceased individuals. Nevertheless, the SSA data used by FSA for this match have been incomplete, and our review of a sample of payments made to deceased individuals raised some questions.

FSA Established Procedures to Compare Its List of Program Participants with a Master List of Deceased Individuals

Since our 2007 report and enactment of the 2008 Farm Bill, FSA has taken steps to implement and strengthen procedures to prevent improper payments to program participants who have died.[10] Specifically, in 2007, FSA began computer matching its payment data with a list of deceased individuals, directed county and state offices to review the results of this match, and established procedures and updated guidance accordingly. According to FSA officials, to identify program participants who had died, in fiscal year 2007 FSA began computer-matching program participants' names, addresses, and Social Security numbers—stored in FSA's primary database of farm program participants—against SSA's death master file. This match identifies any program participants who have died. FSA then compares this list of deceased program participants against its list of participants to whom payments were made during the previous quarter and creates a report listing deceased participants who have received payments. In 2008 FSA began performing these matches quarterly—twice a year more than required under the 2008 Farm Bill; FSA also directed its county and state offices to review these reports to determine whether payments were proper or improper. Every quarter, county officials code each payment to a deceased participant as proper or improper, and state officials review and verify the counties' coding. According to FSA officials, most payments found to be improper are to be recovered by FSA.

By the end of December 2010, FSA had issued a final rule clarifying the regulations governing payments earned by a person who died and describing strengthened procedures for matching program participants with SSA's death master file and having county, then state, offices check which payments to deceased individuals were proper and improper. The agency further improved this process in 2011, by moving from electronic spreadsheets to a web-based system, which, officials told us, greatly improved data accuracy and ease of reviewing and coding payments made to deceased individuals.

In addition to implementing this quarterly data-matching and review process, FSA has revised and updated its handbooks since the 2008 Farm Bill to include guidance related to making payments in cases where a program participant has died. The revised guidance defines the respective roles to be played by county and state offices, clarifies how county offices are to record payments to deceased individuals as proper or improper, and offers guidance for both county and state offices reviewing these records. Agency officials may also take other steps to find out about deceased program participants: in some counties, they may check obituaries in local papers or telephone program participants each year to ask if any changes have occurred among eligible participants.

Overall, according to FSA officials, establishing these procedures has enabled the agency to identify thousands of individuals—17,409 in fiscal year 2011 and 13,684 in fiscal year 2012, for example—who were paid after their dates of death. Of a total of 28,613 deceased individuals who were paid in 2011 and 2012,[11] FSA determined that 1,799 individuals, or about 6 percent, were paid a total of $3.3 million in improper payments during the 2-year period. According to figures from FSA's *DMF Review Report* for fiscal years 2011 and 2012,[12] FSA has recovered approximately $1 million of these improper payments and, according to agency officials, continues to pursue the remaining amount of improper payments.

Incomplete Data and FSA's Payment Review Process Have Hindered FSA's Ability to Identify Improper Payments to the Deceased

The version of SSA's death master file against which FSA matches its payment records has been incomplete. Specifically, FSA has been matching its payment records against the public version of the death master file. This version—containing about 87 million death records and available to the public for purchase—lists all deaths since 1936 that have been reported to SSA by sources other than the states, such as hospitals and funeral homes. According to SSA documentation, to ensure confidentiality under section 205(r) of the Social Security Act, death records provided to the administration by the states are not to be publicly disclosed, except to other agencies that pay federally funded benefits. In May 2013, we testified before the U.S. Senate's Committee on Homeland and Governmental Affairs that according to SSA officials, the complete death master file contains approximately 98 million records, about 11 million more records than the public version of the file.[13] Therefore, until the agency begins matching its payment records to the complete death master file, it may continue to miss deceased individuals to whom it should no longer be making payments.

During our review, FSA took steps to seek access to the complete death master file, and in January 2013, FSA received approval from SSA to obtain such access, along with NRCS and RMA. FSA officials said that they have been coordinating with NRCS and RMA on how they will use, share, and pay for this access. As of early June 2013, however, FSA had not received the complete file from SSA or incorporated these data into its quarterly reviews of payments made to deceased individuals.[14]

In addition, our review of FSA's payment files raised questions about the state and county offices' coding and review of payments as proper or improper. We examined a generalizable random sample of 100 payments that were made to deceased individuals over a 1-year period beginning in April 2011 and coded as proper, and we estimated that FSA county offices coded 91 percent of these payments correctly.[15] On the basis of supporting evidence FSA provided us, we found 9 payments that did not have sufficient support to be coded as proper. For 4 of these 9 payments, the supporting documentation bore the signature of the deceased individual or a representative but was dated after the individual's death. For example, a deceased individual in Alabama received a payment of $4,273 in 2011. County officials coded this payment as proper, but our review identified that some program eligibility documents had been signed and dated in the deceased individual's name more than 6 months after the individual had died. We spoke with the relevant county and state officials about this case, and they agreed that this payment should have been coded as improper. An agency official told us that the deceased individual's heirs submitted correct paperwork for the next fiscal year and that FSA did not attempt to recover the 2011 payment. The remaining 5 payments, all made through one program and paid to deceased individuals after their dates of death, were coded by counties as proper but were actually improper.[16] According to FSA officials, the statutory requirements for this program differ from those of most FSA programs in that they do not permit payments to a deceased individual after the date of death under any circumstances, and any such payment must be recovered. Officials told us that under this program, even if payments go to the correct heirs, such as a spouse, they are still considered improper until the right to receive payment has been transferred to an heir. The officials told us that some counties nevertheless code such payments as proper while awaiting completion of transfer paperwork, even though the payments are actually improper; the officials agreed that the 5 payments we found were indeed improper. According to an FSA official, in May 2012, FSA updated its guidance related to this program, including how to handle payments when a participant dies, and reminded relevant officials of the proper procedures.

Under FSA's current procedures, state offices review the county offices' coding of payments to deceased individuals and confirm whether these payments were proper or improper. Our review, however, found that some improper payments can be overlooked. Some officials we spoke with suggested that, for example, counties did not always have a strong incentive to code payments as improper if the errors were simple errors in paperwork and could be easily corrected. From our random sample, we estimated that an error rate of 9 percent could persist.[17] Under standards for internal control in the federal government,[18] agencies are to engage in ongoing monitoring in the course of normal operations, including in activities such as reconciliations, which could directly test the effectiveness of the quarterly review process.[19] FSA's quarterly review process has largely enabled the agency to identify thousands of individuals who were paid after their dates of death. Nevertheless, if some coded payments were revisited, perhaps annually, to ensure that documentation supported the coding of payments as proper, the error rate could be further lowered, particularly for programs or county offices where errors were previously identified.

NRCS Does Not Have Procedures in Place to Prevent Improper Payments to Deceased Individuals

NRCS does not have procedures to prevent improper payments to deceased individuals, and its ability to verify whether payment recipients are dead or alive is limited. As a result, the agency cannot be certain that payments it made to over a thousand deceased individuals are proper.

According to NRCS officials we spoke with, not all conservation payments to deceased individuals are improper because they may have been made for work performed before the individuals died, or they were associated with easement contracts that became part of the deceased individuals' estates and remained linked to their Social Security numbers. In addition, the officials believe that the risk of the agency's making improper payments to deceased individuals is low, in part because NRCS conservation programs have certain built-in protections that help prevent such payments. The officials also explained that NRCS staff frequently interact with conservation program participants at the sites where projects are located. NRCS officials develop a conservation plan for the site in consultation with participants and typically revisit sites to certify implementation and discuss any deficiencies. As a result, they have regular opportunities to become aware of a participant's death.

Further, NRCS officials commented, counties the agency operates in are often small enough for local officials to know most program participants personally, and thus they would know if a participant died. Moreover, NRCS officials stated that the agency requires that heirs of the deceased notify the agency within 60 days of transferring property to another owner because of death or other reasons.[20] NRCS officials acknowledged, however, that they may be unaware of the death of individuals who receive payments as members of an entity, because officials have less contact with members of an entity than with individual program participants. Thus, NRCS may be unaware that a member of an entity has died if other entity members do not notify the agency.

NRCS's built-in protections are limited, however, because the agency does not systematically verify whether its program participants have died, such as by matching participants' Social Security numbers against SSA's death master file. Under a memorandum of agreement with FSA, such a match is to be performed by FSA, and FSA is to provide NRCS with a list of "current year program payment recipients" who are deceased. FSA officials, however, told us that because they do not know which program participants have been paid by NRCS, they have not provided such a list to NRCS. FSA officials said they are working on a new memorandum of agreement (the previous one expired at the end of fiscal year 2012), in which they hope each agency's responsibilities will be better defined, to enable FSA to provide the matching service. Moreover, FSA and NRCS officials said that the agencies are also coordinating with each other to acquire SSA's complete death master file.

To examine whether NRCS was making potentially improper payments to deceased individuals, we used program participants' Social Security numbers to match program payment data the agency provided to us for fiscal year 2008 through April 2012 against SSA's complete death master file. In so doing, we estimate that NRCS made $10.6 million in payments on behalf of 1,103 individuals 1 year or more after death.[21]

To better understand the reasons for payments that appeared to have been made on behalf of deceased individuals 1 or more years after death, we presented NRCS with six sample cases of such payments, and NRCS explained that four of the six cases were proper payments. For example, in one case, a data entry error appears to have linked the Social Security number of an individual who died in 2001 with multiple program payments made from 2008 through 2012—payments that had not been made to the deceased individual but in actuality, to the appropriate living participants. In another case, NRCS made two direct deposits into the bank account of a participant in a working-lands program 21 months and 32 months after the participant's death. No one had notified the agency of the participant's death or transferred the program contract to his legal heirs, although according to agency officials, the terms of his will did transfer his property to his widow. In another case, however, NRCS did not have proper signatures on a program contract for two payments made in 2008, and agency officials acknowledged that they would not have paid participants had they been aware of this error.[22]

Without procedures such as matching its program participants with the death master file and reviewing those matches, NRCS does not know how many payments it made on behalf of deceased individuals, how often, or in what amounts. Under the standards for internal control in the federal government, agencies are to clearly document internal control in the form of management directives, administrative policies, or operating manuals, and the documentation should be readily available for examination.[23] Furthermore, without reviewing each payment apparently made to a deceased individual, NRCS cannot know whether such payments were proper or improper. Under the standards for internal control in the federal government, monitoring is to be performed continually and include regular management and supervisory activities, comparisons, and reconciliations, which could identify potentially improper payments to deceased individuals. These standards indicate that monitoring should assess the quality of performance over time and ensure that the findings of audits and other reviews are promptly resolved.[24]

RMA Does Not Have Procedures in Place to Prevent Improper Subsidies on Behalf of Deceased Individuals

RMA does not have procedures in place to prevent improper subsidies on behalf of deceased individuals. Specifically, the agency does not systematically verify its policyholders by matching these policyholders' information against SSA's death master file. As a result, RMA may have provided potentially improper crop insurance subsidies and administrative allowances on behalf of thousands of deceased individuals. According to agency officials, however, the crop insurance cycle provides one or more opportunities each year to verify that an individual on whose behalf a subsidy or allowance is paid is alive: policyholders are required to provide a signed certification when filing their annual reports on production and yield and provide a written signature if they file a claim. RMA officials also told us that in 2007—on a one-time basis—the agency compared policyholders with the public death master file and, working with its partner insurance companies, updated and corrected records in its central database of participants, claims, and subsidies, all of which are linked by Social Security number. RMA officials commented that they have not compared the agency's database of policyholders with the public death master file since then because they learned that this file does not have complete information on deceased individuals. Along with FSA and NRCS, however, RMA received approval from SSA for access to the complete file and, according to RMA officials, is coordinating with the other two agencies to begin using it.

To determine whether some of RMA's subsidies and allowances may have been provided on behalf of policyholders who were deceased, we matched policyholders' Social Security numbers from RMA's crop insurance subsidies and administrative allowance data for crop reinsurance years 2008 through 2012 against SSA's complete death master file. We found that approximately $22 million in subsidies and allowances may have been provided on behalf of 3,434 policyholders 2 or more reinsurance years after death.[25] To better understand the reasons for subsidies that appeared to have been made to policyholders after death, we discussed with RMA five sample cases of such subsidies. RMA believed that all were proper, explaining that several were due to errors, such as incorrect Social Security numbers, in RMA's database. Without reviewing each subsidy that could have been made to a deceased individual and reconciling it with SSA's complete death master file, however, RMA cannot know whether such subsidies were proper or improper. Without such reconciliations or matches, RMA is not employing monitoring specified in federal internal control standards.[26]

The Agricultural Risk Protection Act of 2000 requires RMA to use data mining—a technique for extracting knowledge from large volumes of data—to administer and enforce rules for the crop insurance program. The agency uses this technique to (1) develop scenarios of potential program abuse by policyholders, insurance agents, and claims adjusters and (2) query the database that contains crop insurance data and information on weather, soil, and land surveys to generate reports and lists of participating farmers associated with anomalous payments for claims.[27] Since 2001, RMA has worked with a data-mining contractor,[28] which produces reports that, among other results, identify policyholders associated with anomalous payments for claims, which could signal potential cases of fraud, waste, and abuse. Anomalies may include excessive crop yields, frequent filing of claims, and unusually large losses. On finding anomalies, RMA or the insurance company is to conduct further review to determine if fraud, waste, and abuse have in fact occurred, because some anomalous claims, such as those due to unusual weather or other farm conditions, may be legitimate. If, however, claim payments appear in the name of deceased individuals rather than in the names of present, living owners, some anomalous claim patterns might not be visible, and RMA's data-mining contractor could miss patterns that might signal fraud. In addition, because the agency has not had procedures in place to match its claims data against the death master file, we found that it may have unknowingly paid 1,416 policyholders more than $19 million for claims during reinsurance years 2008 through 2012 on behalf of individuals who may be deceased.[29] Under the standards for internal control in the federal government, agencies are to clearly document internal control in the form of management directives, administrative policies, or operating manuals, and the documentation should be readily available for examination.[30] Without accurate data on policyholders, RMA may be unable to rely on results from data mining and therefore be less likely to detect fraudulent, wasteful, or abusive crop insurance claims.

CONCLUSION

Since our July 2007 report and the enactment of the 2008 Farm Bill, FSA has taken important steps to prevent improper payments to deceased individuals. We commend the agency for exceeding the 2008 Farm Bill's requirement that it match its payment records against SSA's death master file twice a year—the agency performs this match quarterly—and for reviewing all payments made to deceased individuals to determine whether they were proper or improper. In addition, FSA, NRCS, and RMA began working together during our review to acquire SSA's complete death master file, an effort we also commend, and they have received approval from SSA for access. FSA's quarterly review process has largely enabled the agency to identify thousands of individuals who were paid after their dates of death. Nevertheless, under current procedures, FSA may not verify all improper payments to deceased individuals, and an error rate of about 9 percent in coding payments could persist. As FSA conducts its reviews, if it employs ongoing monitoring activities, such as reconciliations, to ensure that county offices' coding of payments is supported by documentation, the error rate could be reduced. Furthermore, until and unless NRCS and RMA develop and implement procedures to have their payment or subsidy data records matched against

SSA's complete death master file, either through coordination with FSA or on their own, these agencies cannot know if they are providing payments to, or subsidies on behalf of, deceased individuals; how often they are providing such payments or subsidies; or in what amounts. Without such procedures, NRCS and RMA are not employing internal controls specified in federal standards. And without reviewing each payment to or subsidy provided on behalf of a deceased individual, the agencies do not know if each payment or subsidy is proper or improper. In addition, without accurate information on policyholders, RMA may be unable to rely on results from data mining.

RECOMMENDATIONS FOR EXECUTIVE ACTION

We are making the following three recommendations to the Secretary of Agriculture:

- To further strengthen FSA's procedures for preventing improper payments to deceased individuals, we recommend that the Secretary of Agriculture direct the Administrator of FSA to employ ongoing monitoring activities, such as reconciliations, to ensure that county offices' coding of payments is supported by documentation.
- To help NRCS prevent improper payments to deceased individuals, we recommend that the Secretary of Agriculture direct the Chief of NRCS to develop and implement procedures to prevent potentially improper payments to deceased individuals, including (1) coordinating roles and responsibilities with FSA to ensure that either FSA or NRCS matches NRCS payment files against SSA's complete death master file and (2) reviewing each payment to a deceased individual to ensure that an improper payment was not made.
- To help RMA prevent improper crop insurance subsidies on behalf of deceased individuals and to improve the effectiveness of its data mining, we recommend that the Secretary of Agriculture direct the Administrator of RMA to develop and implement procedures to prevent potentially improper subsidies on behalf of deceased individuals, including (1) matching RMA's crop insurance records against SSA's complete death master file and (2) reviewing each subsidy provided on behalf of a deceased individual to ensure that an improper subsidy was not provided.

AGENCY COMMENTS AND OUR EVALUATION

We provided a draft of this report to USDA for review and comment. USDA provided written comments, which are summarized below. In its comment letter, USDA generally agreed with our report's findings and recommendations but stated that it believes we inaccurately represented NRCS and RMA as having no procedures in place to identify deceased participants. USDA stated that it believes the agencies' normal operating procedures provide opportunities to identify deceased participants. In its comment letter, USDA stated that for NRCS conservation easement programs, transferring property rights involves a title escrow agent, thereby providing an opportunity to determine whether any individuals identified on a deed of trust have died. For RMA, the letter stated that the agency's preliminary analysis of sample cases of deceased individuals that we provided suggests that the potential scope of remaining questionable payments to deceased individuals is more limited than we reported. In addition to the fact that someone is paying the premium each year, the letter stated, the crop insurance cycle provides one or more opportunities each year to verify that an individual on whose behalf a subsidy or allowance is paid is alive—information we have incorporated into our report.

Although NRCS and RMA can identify some deceased participants during their normal operations, we do not believe that identifying deceased individuals during normal operations is a reliable substitute for having a systematic process. As we noted in the report, the agencies have not had specific procedures in place to verify and prevent potentially improper payments to deceased individuals. Indeed, in our review, we found that, without reviewing whether these payments or subsidies were proper, NRCS made payments to more than 1,000 deceased individuals in fiscal year 2008 through April 2012 and that RMA provided subsidies and allowances on behalf of more than 3,000 deceased policyholders in reinsurance years 2008 through 2012. We are therefore pleased to learn from USDA's comment letter that RMA has begun implementing formal, systematic procedures to identify and prevent subsidies on behalf of deceased individuals consistent with our recommendation. According to the comment letter, effective May 1, 2013, RMA implemented a new computer matching procedure to check federal crop insurance program eligibility, subsidies, and payments to policyholders against the public version of the death master file, and, when the complete death master file is available, RMA is prepared to integrate it into the agency's computer matching system. The steps RMA is taking are promising. We encourage NRCS to take similar steps, because until and unless NRCS has its payment data records matched against SSA's complete death master file, it cannot know if it is providing payments to deceased individuals or whether these payments are proper or improper.

Sincerely yours,

Daniel Garcia-Diaz
Director, Natural Resources and Environment

APPENDIX I: OBJECTIVES, SCOPE, AND METHODOLOGY

The objectives of our review were to determine the extent to which procedures are in place to prevent (1) the Farm Service Agency (FSA) from making potentially improper payments to deceased individuals, (2) the Natural Resources Conservation Service (NRCS) from making potentially improper payments to deceased individuals, and (3) the Risk Management Agency (RMA) from providing potentially improper subsidies on behalf of deceased individuals.

To address all three objectives, we reviewed relevant sections of the Food, Conservation, and Energy Act of 2008 (2008 Farm Bill), the Improper Payments Information Act of 2002, the Improper Payments Elimination and Recovery Act of 2010, the Improper Payments Elimination and Recovery Improvement Act of 2012, and the Do Not Pay initiative. We also reviewed relevant studies prepared by the U.S. Department of Agriculture's (USDA) Office of Inspector General, as well as our own past reports.

To determine the extent to which FSA has procedures in place to prevent potentially improper payments to deceased individuals, we reviewed agency guidance, such as FSA's *Handbook on Common Management and Operating Provisions*, 1-CM (Revision 3), as well as FSA's reports compiling the results of its matches of payment data against the Social Security Administration's (SSA) death master file. FSA refers to this report as the *DMF Review Report*. We also interviewed agency officials at FSA's Washington, D.C., headquarters and at its Kansas City, Missouri, information technology office about agency procedures, guidance, payment processes, and the identification of improper payments. In addition, we interviewed FSA officials about the extent to which the agency follows requirements to compare agency data with the data in SSA's death master file and the steps the agency takes to recover improper payments. To obtain information about agency procedures and adherence to guidance, we interviewed FSA's state and county officials in California, Illinois, Kansas, Missouri, and county officials in Texas. We selected offices in Texas, Illinois, and Kansas because these states represent the highest numbers, respectively, of FSA payments made to deceased individuals. We selected the Missouri office because of the state's large number of FSA payments made to deceased individuals and its geographic proximity to other states we visited, and we selected California for geographic diversity.

To determine if state and county offices accurately coded payments made to deceased individuals as proper or improper, we analyzed a generalizable, random sample of payments made by FSA to deceased individuals from April 2011 through March 2012, and we reviewed and analyzed supporting documentation to determine the individuals' eligibility. Specifically, we randomly selected 100 payments made during that 1-year period. The sample size was chosen to provide a margin of error for an attribute measure of no greater than plus or minus 10 percentage points at the 95 percent level of confidence. Because we followed a probability procedure based on random selections, our sample is only one of a large number of samples that we might have drawn. Since each sample could have provided different estimates, we express our confidence in the precision of our particular sample's results as a 95 percent confidence interval (i.e., plus or minus 5 percentage points). This interval would contain the actual population value for 95 percent of the samples we could have drawn. We also randomly selected for case study analysis a nongeneralizable sample of 20 payments made to deceased individuals during the same time period and deemed erroneous by FSA. For the case study analysis, we reviewed and analyzed the supporting payment eligibility documentation of these improper payments. Because this analysis was a case study, the results cannot be generalized to all erroneous payments identified by FSA, but they provide examples of the kinds and types of improper payments made to deceased individuals. We assessed the reliability of FSA's payment data by (1) reviewing existing information about the data and the system that produced them and (2) interviewing agency officials knowledgeable about the data. We determined that the data were sufficiently reliable for the purposes of our review.

To determine the extent to which NRCS has procedures in place to prevent potentially improper payments to deceased individuals, we interviewed agency officials at NRCS's Washington, D.C., headquarters office and state and county offices in California, Illinois, Kansas, and Missouri. We selected these offices for geographic diversity and because of their close proximity to meetings we were holding with FSA officials. Using Social Security numbers, we compared the death master file with NRCS payment data for conservation programs from fiscal year 2008 through April 2012 to determine the number individuals paid by NRCS who died 1 year or more before the payment dates and the amount of these potentially improper payments. We assessed the reliability of SSA's death master file by (1) performing electronic testing of required data elements and (2) reviewing relevant documentation. We determined that the data were sufficiently reliable for the purposes of our review. We included the following conservation programs in our analysis: the Agricultural Water Enhancement Program, Agricultural Management Assistance, Chesapeake Bay Watershed Initiative, Conservation Security Program, Conservation Stewardship Program, Environmental Quality Incentives Program, Farm and Ranch Lands Protection Program, Grassland Reserve Program, Healthy Forests Reserve Program, Wetlands Reserve Program, and Wildlife Habitat Incentives Program. We did not review a portion of payments made under these programs because NRCS was unable to provide us with the Social Security numbers for some program participants. We assessed the reliability of NRCS's data by (1) performing electronic testing of required data elements and (2) interviewing agency officials knowledgeable about the data. We determined that the data were sufficiently reliable for the purposes of our review.

To determine the extent to which RMA has procedures in place to prevent improper subsidies to deceased policyholders, we interviewed RMA officials to obtain information about the steps the agency takes to verify its records of eligible policyholders. To determine the extent to which RMA's records are regularly updated to verify whether policyholders have died, we compared the death master file with crop insurance data from RMA for reinsurance year 2008 through reinsurance year 2012 and produced a list of premium subsidies, administrative allowances, and claims payments provided on behalf of policyholders 2 or more reinsurance years after death. As mentioned earlier, we assessed the reliability of SSA's death master file and determined that the data were sufficiently reliable for the purposes of our review. We assessed the reliability of RMA's crop insurance data by (1) reviewing related documentation, (2) performing electronic testing of required data elements, and (3) interviewing agency officials knowledgeable about the data. We determined that the data were sufficiently reliable for the purposes of our review.

We conducted this performance audit from March 2012 through June 2013 in accordance with generally accepted government auditing standards. Those standards require that we plan and perform the audit to obtain sufficient, appropriate evidence to provide a reasonable basis for our findings and conclusions based on our audit objectives. We believe that the evidence obtained provides a reasonable basis for our findings and conclusions based on our audit objectives.

End Notes

[1] GAO, *Federal Farm Programs: USDA Needs to Strengthen Controls to Prevent Improper Payments to Estates and Deceased Individuals*, GAO-07-818 (Washington, D.C.: July 9, 2007); *Federal Farm Programs: USDA Needs to Strengthen Controls to Prevent Payments to Individuals Who Exceed Income Eligibility Limits*, GAO-09-67 (Washington, D.C.: Oct. 24, 2008); *Crop Insurance: Savings Would Result from Program Changes and Greater Use of Data Mining*, GAO-12-256 (Washington, D.C.: Mar. 13, 2012); and *USDA Office of Inspector General, U.S. Department of Agriculture Improper Payments Elimination and Recovery Act of 2010 Compliance Review for Fiscal Year 2012*, Audit Report 50024-0004-11 (Washington, D.C.: Mar. 14, 2013).

[2] GAO

[3] GAO, *Farm Bill: Issues to Consider for Reauthorization*, GAO-12-338SP (Washington, D.C.: Apr. 24, 2012).

[4] Pub. L. No. 107-300, 116 Stat. 2350, amended by Improper Payments Elimination and Recovery Improvement Act of 2012, Pub. L. No. 112-248, 126 Stat. 2390 (2013), and Improper Payments Elimination and Recovery Act of 2010, Pub. L. No. 111-204, 124 Stat. 2224.

[5] Pub. L. No. 112-248 § 5, 126 Stat. 2390, 2392 (2013).

[6] The death master file is a national database of deceased individuals who had Social Security numbers and whose deaths were reported to SSA. It contains information on date of birth, date of death, and state or country of residence for each decedent and is a tool for identifying deceased individuals in a timely way.

[7] A crop reinsurance year begins on July 1 and ends on June 30.

[8] USDA officials said that since 2009, the agency has assigned payments on behalf of an estate directly to the estate's heirs, eliminating incentives for heirs to keep an estate open to receive payments in excess of their own payment limits.

[9] For purposes of this report, we consider the following NRCS programs to be working- lands programs: Agricultural Water Enhancement Program, Agricultural Management Assistance, Chesapeake Bay Watershed Initiative, Conservation Security Program, Conservation Stewardship Program, Environmental Quality Incentives Program, and Wildlife Habitat Incentives Program. We consider the following NRCS programs to be easement programs: Farm and Ranch Lands Protection Program, Grassland Reserve Program, Healthy Forests Reserve Program, and Wetlands Reserve Program.

[10] FSA considers a payment to a deceased individual proper if the payment was earned before death and the individual is otherwise eligible. It considers a payment to a deceased individual erroneous if the individual did not meet program eligibility requirements, such as the requirement to be actively engaged in farming. For purposes of this report, we refer to these erroneous payments as improper.

[11] The total number of unique individuals to whom payments were made in 2011 and 2012 (28,613) does not equal the sum of the number of individuals to whom payments were made in 2011 (17,409) and 2012 (13,684) because a single individual may have been paid in both 2011 and 2012.

[12] FSA's *DMF Review Report* is a report compiling the results of its matches of payment data against SSA's death master file.

[13] GAO, *Social Security Administration: Preliminary Observations on the Death Master File*, GAO-13-574T (Washington, D.C.: May 8, 2013).

[14] FSA officials told us that FSA has not yet received the complete death master file because FSA is awaiting SSA approval for a fourth USDA agency—National Agricultural Statistics Service—to access this file.

[15] The 95 percent confidence interval associated with the estimated FSA accuracy rate of 91 percent for payments that were made to deceased individuals and coded as proper is (84, 96).

[16] These payments were made through the Tobacco Transition Payment Program.

[17] The 95 percent confidence interval for this estimate is (4,16).

[18] GAO, *Standards for Internal Control in the Federal Government*, GAO/AIMD-00-21.3.1 (Washington, D.C.: 1999).

[19] According to the standards, the scope and frequency of such direct testing should depend primarily on the assessment of risks and the effectiveness of ongoing monitoring procedures.

[20] This requirement is stated in the terms and conditions of the conservation program contract signed by participants in NRCS's conservation programs.

[21] NRCS officials told us that they immediately update their records when they learn about a participant's death. We selected a 1-year time frame to provide sufficient time for the reporting and updating of death records.

[22] NRCS's conservation program contract requires signatures from program participants (owners, operators, or tenants). The signature of a person with an approved power of attorney may also be acceptable.

[23] GAO

[24] GAO

[25] We found 21 agents and 2 loss adjusters whose Social Security numbers matched Social Security numbers in the death master file. We chose 2 reinsurance years as our unit of analysis to provide sufficient time for the reporting and recording of new deaths.

[26] GAO

[27] The Agricultural Risk Protection Act of 2000, Pub. L. No. 106-224, 114 Stat. 358, amended the Federal Crop Insurance Act. We have evaluated RMA's data-mining activities in several reports: see GAOGAO, *Crop Insurance: Continuing Efforts Are Needed to Improve Program Integrity and Ensure Program Costs Are Reasonable*, GAO-07-944T (Washington, D.C.: June 7, 2007); *Crop Insurance: More Needs to Be Done to Reduce Program's Vulnerability to Fraud, Waste, and Abuse*, GAO-06-878T (Washington, D.C.: June 15, 2006); and *Crop Insurance: Actions Needed to Reduce Program's Vulnerability to Fraud, Waste, and Abuse*, GAO-05-528 (Washington, D.C.: Sept. 30, 2005).

[28] RMA's data-mining contractor is the Center for Agribusiness Excellence at Tarleton State University, Stephenville, Texas.

[29] We calculated this $19 million by first matching policyholders' Social Security numbers in RMA's crop insurance claims payment data from this period against SSA's complete death master file and then summing the dollar amounts of each payment to a policyholder who, according the death master file, had been deceased for 2 or more reinsurance years.

[30] GAO

INDEX

A

B

C

D

E

F

G

H

Q

R

S

T

U

W

Y